Flame Resistance Characteristics and Its Influence Mechanism

Zhirong Wang · Xingyan Cao

Flame Resistance Characteristics and Its Influence Mechanism

Springer

Zhirong Wang
College of Emergency Management
Nanjing Tech University
Nanjing, China

Xingyan Cao
College of Safety Science and Engineering
Nanjing Tech University
Nanjing, China

The authors are grateful for National project funding for Key R & D programs under Grant No. 2023YFC3009900, and Prof. Zhirong Wang and Prof. Xingyang Cao are Research Fellows at the Institute of Emergency Governance and Policy, Nanjing Tech University, China. This work was supported by The Open Fund of the State Key Laboratory of Fire Science (SKLFS) Program (HZ2025-KF09) and National Natural Science Foundation of China (52274210; 51806056).

ISBN 978-981-95-6118-6 ISBN 978-981-95-6119-3 (eBook)
https://doi.org/10.1007/978-981-95-6119-3

This Springer imprint is published by the registered company Springer Nature Singapore Pte Ltd.
The registered company address is: 152 Beach Road, #21-01/04 Gateway East, Singapore 189721, Singapore

Competing Interests The authors have no competing interests to declare that are relevant to the content of this manuscript.

Contents

Chapter 1
Introduction

The transportation of hydrogen has always been a difficult problem restricting the development of hydrogen energy industry [1, 2]. As a highly active flammable gas, it is inevitable that hydrogen leakage occurs due to the multiple uncertain factors during its storage and transportation, which is easy to cause serious gas explosion accident. The flame arrester, as an explosion-retardant device, can effectively prevent the propagation of high activity flammable gas explosion flame and is widely utilized in gas transportation pipeline of petrochemical enterprises [3, 4]. As a key component for explosion prevention, its failure will lead to incalculable consequences. How to effectively protect hydrogen doped gas explosion is currently a hot and difficult research topic [5].

Wire mesh flame arrester was considered an effective flame-retardant deceives. Relevant research on the flame resistance of wire mesh on combustible gas explosion has been concerned [6, 7]. Sun et al. [8] found out that it had an excellent fire-preventing effect compared with the foam ceramics. Chen et al. [9] researched the fire-preventing effect of wire mesh on gas explosion flame inside the closed pipeline. According to the result, it pointed out that the pressure could be decreased linearly with increasing of pore density. Jin et al. [10, 11] discussed its flame resistance effect on hydrogen explosion and proposed that the explosion intensity of highly active combustible gas could be also evidently decreased. In addition, the difficulty of flame resistance was related to the distance between the wire mesh and ignition source and was enlarged as the distance increased. Cui et al. [12] proposed that the layer and mesh numbers of wire mesh as the important influence parameters had an obviously impact on its flame resistance effect and determined the influence of its installation position on explosion inhibition. Lu et al. [13] and Wang et al. [14] researched the influence of wire mesh structure parameter on venting length of flame, and proposed that the length could be continuously reduced as the layer number increased. And the explosion parameter was also continuously reduced. Under the condition of 5 layers and 60 meshes, the flame could be completely prevented. Besides, Zhang et al. [15] researched the flame resistance effect of wire mesh inside the connected vessels,

Z. Wang and X. Cao, *Flame Resistance Characteristics and Its Influence Mechanism*,
https://doi.org/10.1007/978-981-95-6119-3_1

and pointed out that the wire mesh could protect the adjacent vessel by effectively preventing flame propagation. Compared with the mesh number, the layer number of wire mesh presented the more significant impact on the pressure. Cheng et al. [16] discussed the combined action of obstacle and wire mesh on the flame propagation process of methane explosion. Under the action of both, the flame could be not effectively prevented due to its acceleration propagation after the obstacle existed. However, an opposite inhibition result was presented as only the wire mesh existed. Wang et al. [17] studied the combined action of the sidewall discharge and wire mesh flame resistance on the methane explosion. Compared with the single wire mesh, the synergistic effect had a more obvious impact on the pressure. And the wire mesh should be installed at a proper position according to the change of explosion parameter. Duan et al. [18] analyzed the flame resistance effect of porous material on methane/air explosion, and proposed that the inhibition effect mainly depended on the pore diameter and its thickness. Compared with the latter, the effect of pore diameter on the pressure rising rate was more significant. In addition, the resistance of flame during propagation inside the porous material was enlarged and its propagation velocity was evidently diminished as the thickness was increased.

Besides, Mohammed et al. [19] evaluated the effectiveness of corrugated flame arrester in reducing the intensity of methane explosion and pointed out that the propagation velocity of flame and intensity of pressure wave could be effectively attenuated. Mendiburu et al. [20] proposed that the corrugated flame arrester had the best quenching ability and the pressure drop was smallest by comparing the multiple types of flame arresters. And the flame resistance efficiency was related to the size of narrow channel, the thickness of flame-retardant element and the velocity of flame propagation. Yue et al. [21] and Kakutkina et al. [22] pointed out that the flame resistance effect could be evidently improved with the decrease of porosity and the increase of flame-retardant element thickness. Sun [23] proposed that the length of flame quenching presented the exponential increase as the porosity enlarged, but the pressure and flame-retardant velocity at the front end of flame arrester decreased as the quadratic and exponential functions respectively. As the flame-retardant element thickness increased, the probability of flame resistance success was obviously improved and the velocity was also greatly decreased. Wan et al. [24] found that the initial pressure had an obvious impact on the explosion resistance effect. And it could penetrate the flame arrester to cause the failure occurrence as the initial pressure was higher than its rated pressure. Palmer and Tonkin [25] found that the ripple height was also a key factor for the flame resistance success, and the failure probability was increased as its enlarged. Li et al. [26] proposed that the flame was difficult to quench with the increases of its propagation velocity and the porosity. Besides, the high-speed turbulent flame intensified the exchanges of heat and mass, resulting in the reduction of heat conduction time and making it difficult to quench. Besides, the velocity of flame propagation was also related to gas composition, the ignition position and the pipeline structure.

Currently, scholars had conducted relevant research on the explosion resistance effect and its influence factor. And it mainly focused on the quenching results and influence laws of multiple flame arresters. Nevertheless, the research on the flame propagation characteristics (including the flame structure and its velocity) at the front and rear-ends affected by flame-retardant system during the explosion resistance and the correlation with explosion parameters was relatively rare through the visualization experiment. Due to the involvement of complex fluid flow, the change in the dynamic characteristics of explosion resistance process and the impact pattern was more difficult to reveal. Grasping the morphological transformation of high-temperature flame and change law of pressure at the front and rear ends affected by flame-retardant system could provide the theoretical support for revealing the attenuation law and explosion resistance mechanism of the coupling between the flame and pressure wave inside the narrow channel of corrugated flame arrester. By the visualization experiment, the influence laws of flame resistance characteristics and explosion parameters at the front and rear ends of corrugated flame-retardant system were clarified. Based on the experimental result, the theoretical basis could be provided for the establishment of a calculation model for the explosion resistance parameters of hydrogen related flame retardant system under the influences of multiple factors.

References

1. Zhu YC, Liu G, Cui ZY, Yang HW, Liu F, Jiang BL, Chen L. Effect and mechanism of ionic liquid-polymer composite coating on enhancing hydrogen embrittlement resistance of X80 pipeline steel for hydrogen blended natural gas transportation. Int J Hydrogen Energy. 2024;80:1305–16.
2. Yu QY, Hao YS, Ali K, Hua QS, Sun H. Techno-economic analysis of hydrogen pipeline network in China based on levelized cost of transportation. Energ Convers Manag. 2024;301:118025.
3. Nie ZH, Gao W, Jiang HP, Zhao FY. Quenching characteristics and mechanism of hydrogen-air mixtures by corrugated plate flame arrester under inert conditions. Fuel. 2024;362:130822.
4. Shen YX, Liu XM, Li BB, Zhang C, Liu LL, Li W. Study on flame quenching characteristics inside the Tesla valve structure flame arrester unit. J Loss Prev Process Ind. 2024;90:105351.
5. Wang JZ, Cao XY, Wang ZR, Lin CD, Shen YN. Effect of obstacle parameters on explosion resistance performance of hydrogen crimped ribbon flame arrester. Int J Hydrogen Energy. 2023;48:34440–53.
6. Wen X, Guo Z, Wang F, Pan R, Liu Z, Zhang X. Experimental study on the quenching process of methane/air deflagration flame with porous media. J Loss Prevent Process. 2020;65:104121.
7. Yu MG, Wang XY, Zheng K, Han SX, Chen CD, Si RJ, Wang L. Investigation of methane/air explosion suppression by modified montmorillonite inhibitor. Process Saf Environ Prot. 2020;144:337–48.
8. Sun JH, Zhao Y, Wei CR, Xie S, Huang DH. The comparative experimental study of the porous materials suppressing the gas explosion. Proced Eng. 2011;26:954–60.
9. Chen P, Huang FJ, Sun YD, Chen XX. Effects of metal foam meshes on premixed methane-air flame propagation in the closed duct. J Loss Prev Process Ind. 2017;47:22–8.
10. Jin KQ, Duan QL, Chen JY, Liew KM, Gong L, Sun JH. Experimental study on the influence of multi-layer wire mesh on dynamics of premixed hydrogen-air flame propagation in a closed duct. Int J Hydrogen Energy. 2017;42(21):14809–20.
11. Jin KQ, Wang QS, Duan QL, Chen JY, Sun JH. Effect of metal wire mesh on premixed H_2/air flame quenching behaviors in a closed tube. Process Saf Environ. 2021;146:770–8.

12. Cui YY, Wang ZR, Zhou KB, Ma LS, Liu MH, Jiang JC. Effect of wire mesh on double-suppression of CH_4/air mixture explosions in a spherical vessel connected to pipelines. J Loss Prev Process Ind. 2017;45:69–77.
13. Lu YW, Wang ZR, Cao XY, Cui YY, Sun PP, Qian CJ. Interaction mechanism of wire mesh inhibition and ducted venting on methane explosion. Fuel. 2021;304:121343.
14. Wang ZR, Lu YW, Cao XY, Yu Y, Jiang JC, Jiao F, Ma SC. Wire-mesh inhibition of jet fire induced by explosion venting. J Loss Prev Process Ind. 2021;70:104408.
15. Zhang SF, Wang ZR, Zuo QQ, Jiang JC, Cheng CD. Suppression effect of explosion in linked spherical vessels and pipelines impacted by wire-mesh structure. Process Saf Prog. 2016;35(1):68–75.
16. Cheng FM, Lu JJ, Li TT, Luo ZM. Study on the influence of obstacle on the flame behaviors of methane/air in a semi-confined pipe with wire mesh installed. J Loss Prev Process Ind. 2022;78:104826.
17. Wang Z, Zhang ZL, Tian W, Wang ZS. Coupling effect of side explosion vent and wire mesh on suppressing methane explosion in a closed duct. J Loss Prev Process Ind. 2022;76:104738.
18. Duan YL, Wang S, Yang YL, Li YB, Zheng K. Experimental study on methane explosion characteristics with different types of porous media. J Loss Prev Process Ind. 2021;69:104370.
19. Mohammed J, Ajrash AA, Jafar Z, Behdad M. Application of flame arrester in mitigation of explosion and flame deflagration of ventilation air methane. Fuel. 2019;257:115985.
20. Mendiburu AZ, Norabuena LP, Martinez VA, Silva RQD, Chumpitaz CR. Flame propagation behavior inside a closed duct upstream of a flame arrester housing. Int J Hydrogen Energy. 2024;78:294–302.
21. Yue JP, Wei L, Liu H, Guo S. A novel detonation arrester containing a large disk with long triangular slits: Design and numerical simulation. Process Saf Prog. 2020;10:1–16.
22. Kakutkina NA, Korzhavin AA, Rychkov AD. Burning-through of porous flame arresters with a channel flame-arrester element. Combust Explo Shock. 2009;45:266–73.
23. Sun SC. Research of explosion resistant characteristic in In-line crimped-ribbon flame arrester. Dalian: Dalian University of Technology; 2017.
24. Wan Y, Wang CJ, Li Q, Luo XJ. Experimental study of premixed hydrogen-air flame quenching in a channel with the perforated plate. Fuel. 2020;263:116733.
25. Palmer KN, Tonkin PS. The quenching of flames of various fuels in narrow apertures. Combust Flame. 1963;7:121–7.
26. Li YY, Jiang RP, Li ZP, Xu S, Pan F, Xie LF. Detonation performance and flame quenching characteristics of C_2H_4/N_2O premixed gases. Chin J High Press Phys. 2020;34:1–9.

Part I
Wire Mesh Flame Arrester

Chapter 2
Research on the Flame Resistance Characteristics of Wire Mesh for Syngas Explosion

2.1 Experimental Apparatus and Procedures

The diagram of experimental apparatus for inhibiting syngas explosion by wire mesh is shown in Fig. 2.1. The main structure has been described in in the previous work. In this experiment, self-designed wire mesh flame resistance system [the effective area (A) = 100 mm × 100 mm] was adopted and installed inside the vessel by the support rod to prevent the flame propagation. Its main structure included four support rods with 5 mm diameter, two support plates with 3 mm thickness and the multi-layer wire mesh (see in Fig. 2.1). The multi-layer wire mesh was compressed by the two support plates and the pre-tightening force was provided by bolts. To improve the fixed strength of support plate, two stiffeners with 5 mm width were uniformity set in the transverse and longitudinal directions of the opening area. The distance between the flame resistance system and vessel bottom end was 650 mm. Based on the previous works, the flame could fully develop in this length of vessel. Besides, two thermocouples and pressure transmitters were correspondingly installed in the wall surface of flame resistance system two ends [the height (h) = 50 mm]. By changing the syngas concentration and wire mesh layer number, the flame resistance mechanism was analyzed in-depth [1].

Syngas and air were directly mixed inside the vessel. Moreover, the high voltage with 8 kV was adopted to ignite the mixture gas and the discharge lasted 100 ms. The explosion pressure was gathered by the pressure transmitter with 50 kHz acquisition frequency. Besides, the S-type thermocouple was adopted to record the change of flame temperature with time. A high-speed camera with 4000 fps acquisition frequency was adopted to record the propagation process of syngas explosion flame under different flame resistance conditions. A high-frequency data acquisition card with 2 MHz per channel was applied to realize the electrode discharge and data acquisition.

Before the experiment, the flame resistance system with determined layer and mesh numbers was installed inside the vessel. The mixed gases with four kinds of

Z. Wang and X. Cao, *Flame Resistance Characteristics and Its Influence Mechanism*,
https://doi.org/10.1007/978-981-95-6119-3_2

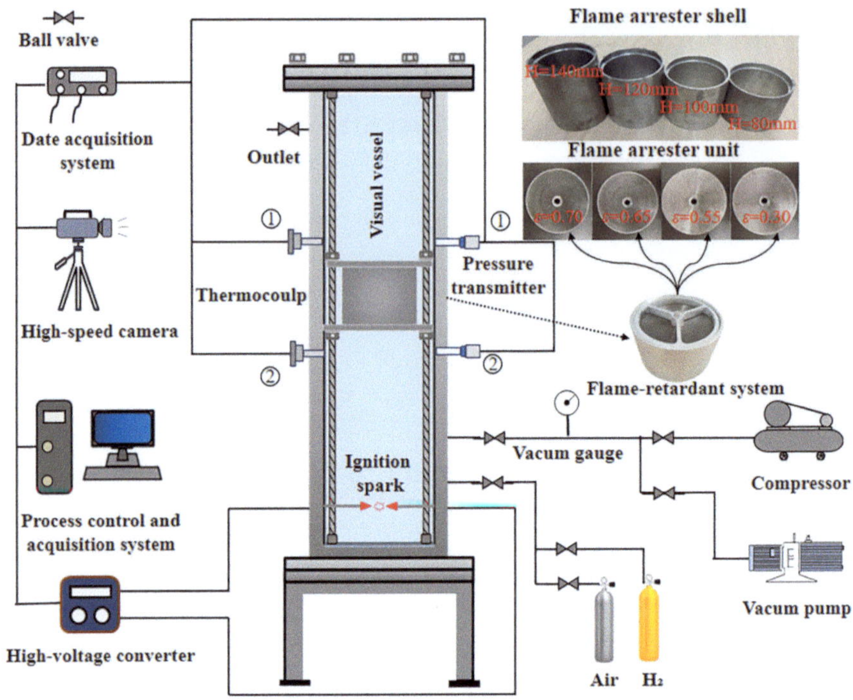

Fig. 2.1 Structure diagram of experimental apparatus

syngas concentrations ($c = 30, 25, 20$ and 15%) were prepared and the initial pressure was 0.1 MPa as the mixture was ignited. After ignition, the flame developed around the ignition electrodes and propagated from the bottom end to top end of vessel. The pressure, flame shape and temperature were evidently affected after adding wire mesh. In order to ensure the validity of data, the experiment under same flame resistance condition was repeated 3 times at least.

2.2 Results and Discussions

2.2.1 Analysis of Flame Resistance Characteristics

To better describe the change of pressure rise with flame propagation during the flame resistance, Fig. 2.2a presents their corresponding relationship under no wire mesh ($c = 30\%$). An obvious corresponding relationship was presented at the support plate two ends. After ignition, the light blue flame developed in a "spherical" structure [2]. Due to the wall surface constraint, the flame acceleratingly propagated upward with a "finger" structure. And the velocity history was obviously increased. As $t = 8.75$ ms, the pressure below support plate started to rise. After 11.25 ms, the velocity history appeared an evident decrease trend resulting from the ceaseless increase of flame front curvature radius and the decrease of its surface area with time [3]. However, the support plate as an obstacle changed the structure and area of flame front. The flame near the wall surface was compressed and blocked, resulting in the acceleration propagation of central area flame after it bypassed the support plate (from 62.3 to 282.1 m/s). Meanwhile, the pressure at the support plate upper end started to rise at 9.5 ms. And the history also showed an evident increase trend with the acceleration of flame. As the flame got the brightest, the (dP/dt) history appeared the peak value ($t = 18.75$ ms).

Compared with the support plate lower end, the ΔP_{max} and $(dP/dt)_{max}$ of its upper end were larger. And it was increased respectively from 0.58 to 0.81 MPa and from 329.1 to 449.5 MPa/s. Besides, the reflected pressure wave would propagate downward through the support plate, resulting the secondary acceleration rise of pressure history at its lower end. Hence, the appearance moments of ΔP_{max} and $(dP/dt)_{max}$ were slightly delayed compared with the upper end. The twice acceleration rises of pressure resulted from the self-acceleration propagation of flame and the reverse action of pressure wave [4]. The interval time of (dP/dt) history two peaks was 1.25 ms. Compare with lower end, the flame temperature of upper end was also slightly higher. But the decrease rate of pressure was more significant and the (dP/dt) history appeared a larger negative value [$(dP/dt)_{min} = -148.3$ MPa/s]. This explains that the explosion flow field appeared an evident change after being affected by the obstacle, and the impact on its upper end was more remarkable.

The existence of wire mesh could indirectly affect the pressure rise by affecting the flame propagation. Figure 2.2b presents their corresponding relationship under the flame resistance failure ($c = 30\%$, m = 60, n = 15). At the early stage, the flame presented the obvious "spherical" and "finger" structures. And the acceleration propagation of flame caused the pressure below wire mesh to rise. But the moment for pressure started to rise was slightly advanced ($t = 7.25$ ms). It was because that the reverse action of pressure wave due to the obstruction of wire mesh, further disturbed the flame front and accelerated its propagation [5]. Compared with no wire mesh condition, the velocity appeared a slightly increase trend (from 47.3 to 54.5 m/s). After that, its history experienced a downward process and the reduction extent were more obviously under the flame resistance failure condition. As the flame approached wire mesh, an obvious acceleration process also appeared in the central area of flame front although the existence of wire mesh.

Moreover, the mixed gas above wire mesh was ignited after the flame occurred a stagnation inside the wire mesh ($\Delta t = 4.75$ ms). As the flame accelerated, the pressure above wire mesh was constantly increased due to the enhancement of flame front turbulent. However, the velocity was reduced compared with no wire mesh ($v_{max} = 112.3$ m/s). The flame got the brightest and the peak value of (dP/dt) history appeared at 24.00 ms. However, the ΔP_{max} and $(dP/dt)_{max}$ were obviously reduced. At this stage, the pressure below wire mesh presented the once more accelerated rising process ($t = 24.00$–26.00 ms), but the explosion intensity was decreased [$\Delta P_{max} =$ 0.39 MPa; $(dP/dt)_{max} = 56.6$ MPa/s]. In addition, the interval time of (dP/dt) history

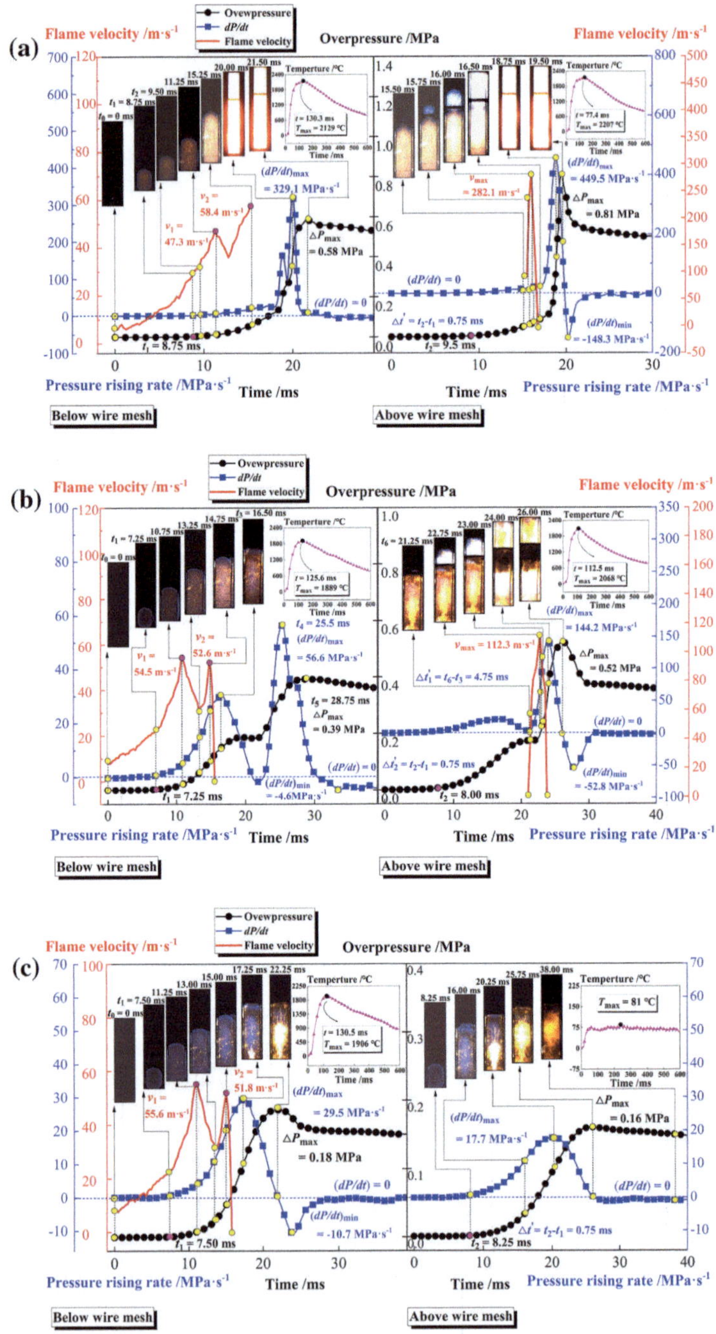

Fig. 2.2 **a** Change of pressure rise with flame propagation under no wire mesh condition ($c = 30\%$). **b** Change of pressure rise with flame propagation under the flame resistance failure condition ($c = 30\%$, $m = 60$, $n = 15$). **c** Change of pressure rise with flame propagation under the flame resistance success condition ($c = 30\%$, $m = 60$, $n = 25$)

two peaks was increased. It explains that the intensity of syngas explosion could be greatly weakened after adding wire mesh. Furthermore, the reflected pressure wave would pass through the multi-layer wire mesh and act on the flame front of lower end, leading to the downward propagation of flame below wire mesh and a rapid reduction of pressure above wire mesh [$(dP/dt)_{min} = -52.8$ MPa/s]. Besides, the existence of wire mesh caused the obvious increase in the flame propagation velocity, further resulted in a slight increase of temperature at the wire mesh upper end (from 1889 to 2068 °C).

Figure 2.2c presents the change of pressure rise with flame propagation under flame resistance success condition ($c = 30\%$, m = 60, n = 25). Both of them also showed an evident corresponding relationship at the wire mesh two ends. Meanwhile, it clear that the flame structure showed a similar change trend to that of flame resistance failure in the initial stage. The acceleration propagation of flame led to an obvious rise of pressure. As the flame approached wire mesh, it also appeared an obvious acceleration process ($t = 15.00$–17.25 ms). As the flame at the lower end got brightest, the (dP/dt) history reached the peak value. However, the flame emerged a stagnation phenomenon after contacting with wire mesh and extinguished inside the flame resistance system. Hence, (dP/dt) history was decreased and appeared a large negative value [$(dP/dt)_{min} = -10.7$ MPa/s]. Especially, the reductions of ΔP_{max} and $(dP/dt)_{max}$ were more significant. For the wire mesh upper end, the partial compression wave could still pass through wire mesh although the flame was extinguished, further causing an obvious increase of pressure. But the ΔP_{max} and $(dP/dt)_{max}$ of upper end were reduced respectively from 0.52 to 0.16 MPa and from 144.2 to 17.7 MPa/s.

Besides, the flame temperature ($t_{max} = 81$ °C) was far below the ignition point of syngas (570 °C). A strong reflected pressure wave would not be generated and pass through the wire mesh to act on the flame of lower end. Hence, (dP/dt) history also did not present the twice rise processes. Nevertheless, the flame temperature was slightly higher than that of the flame resistance failure. Especially, it was evidently decreased compared with the condition without wire mesh.

2.2.2 *Comparison of Flame Propagation*

The comparison of flame propagation processes under different flame resistance conditions ($c = 30\%$) is shown in Fig. 2.3. The velocity history can be obtained according to the alteration of flame front position and is clearly presented in Fig. 2.4. Combined with Figs. 2.3 and 2.4, the flame structure presented the similar change trend in the early stage and changed from the "spherical" to "finger" structure. Meanwhile, the flame experienced three times acceleration processes: the self-accelerating propagation, mesh-accelerating propagation and turbulence-accelerating propagation processes [6]. Restrained by the wall surface, the flame presented the first acceleration propagation process. Thereafter, its curvature radius was increased and surface area was decreased as the flame propagated, causing a significant reduction of velocity history. At this stage, the velocity history was slightly higher after adding wire mesh due to the disturbance of reflected pressure wave resulted from wire mesh on flame front. As the flame closed to flame resistance system, the local pressure near the wall surface increased sharply due to the existence of support plate, resulting in a secondary acceleration propagation of central local flame. Subsequently, the flame emerged the third acceleration process after it passed through wire mesh. And the flame acceleration was more obvious under no wire mesh, followed by the flame resistance failure.

Figure 2.5 illustrates the effect of syngas concentration on the velocity history under the flame resistance condition (c = 15, 20, 25, 30%; n = 15; m = 60). The velocity history and its slope were evidently increased with increasing syngas concentration. It clear that the first peak of velocity history (v_{a1}) presented a rise trend from 25.4 to 54.5 m/s and the corresponding moment was advanced successively. The secondary acceleration process all was evidently appeared as the flame approached wire mesh. Besides, the secondary peak of velocity history (v_{a2}) also showed a change trend (from 23.2 to 52.6 m/s). In particular, their corresponding moments were advanced (from 24.75 to 14.75 ms). As c = 20, 25 and 30%, the flame emerged the third acceleration process, which meant the flame resistance failure. As c = 15%,

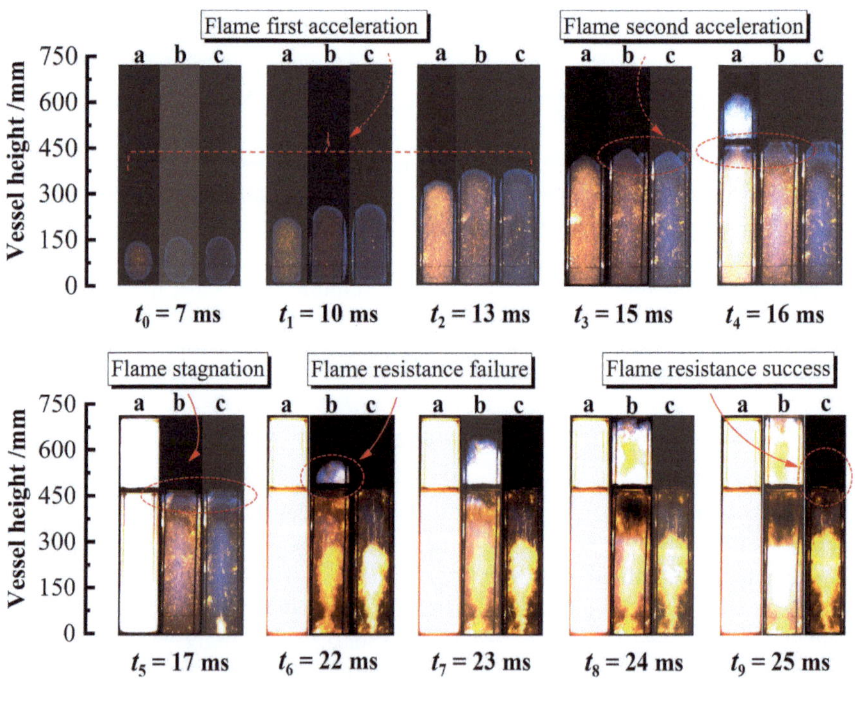

Fig. 2.3 Comparison of flame propagation processes under different flame resistance conditions (c = 30%; **a** No wire mesh, **b** Flame resistance failure, **c** Flame resistance success)

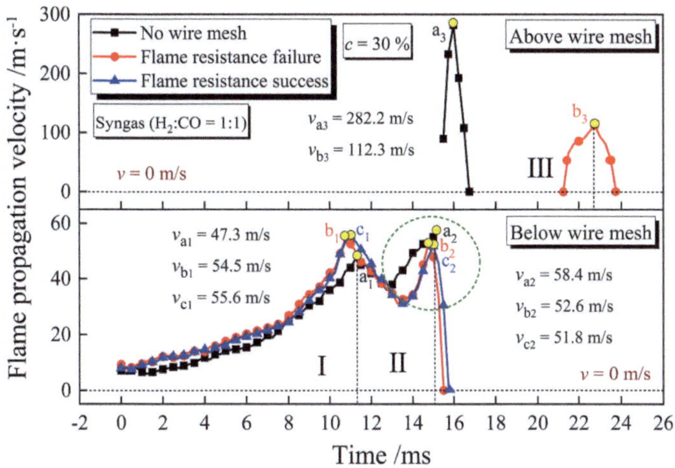

Fig. 2.4 Comparison of velocity histories under different flame resistance conditions ($c = 30\%$)

the flame was effectively prevented and only presented twice acceleration propagation processes. It also explains that the flame resistance effect of wire mesh was related to its propagation velocity. Besides, the flame acceleration was also more significant with increasing syngas concentration as the flame resistance failed (from 73.2 to 112.3 m/s).

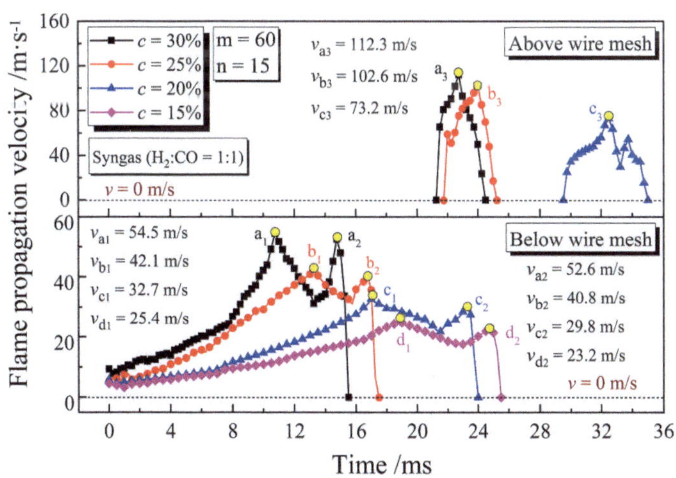

Fig. 2.5 Comparison of velocity histories under different syngas concentrations ($c = 15$, 20, 25 and 30%; n = 15; m = 60)

2.2.3 Analysis of Explosion Pressure

Figure 2.6 shows the comparison of pressure histories under different flame resistance conditions ($c = 30\%$). After adding wire mesh, the explosion intensity could be greatly decreased. Especially for the ΔP_{max}. As the flame resistance failed, it was effectively decreased from 0.81 to 0.52 MPa above wire mesh and from 0.58 to 0.39 MPa below wire mesh. Under flame resistance success, the reduction extent of pressure was more significant (from 0.81 to 0.16 MPa and from 0.58 to 0.18 MPa). Meanwhile, the pressure above wire mesh was higher and the appearance moment of ΔP_{max} was also advanced under flame resistance failure condition. Especially for no wire mesh. However, it can be clearly seen that an opposite result was shown under the flame resistance success condition.

Figure 2.7 illustrates the effect of syngas concentration on pressure histories of wire mesh two ends ($c = 15, 20, 25, 30\%$; n = 60; m = 15). The moment for pressure started to rise was delayed respectively from 7.75 to 28.00 ms and from 8.00 to 27.75 ms with decreasing syngas concentration. Meanwhile, the ΔP_{max} was also successively decreased from 0.52 to 0.14 MPa (the upper end) and from 0.39 to 0.13 MPa (the lower end). In particular, their appearance moments were also evidently prolonged. It indicates that the syngas concentration had an evident effect on the pressure of flame resistance system two ends. Besides, the pressure of upper end was higher compare with the lower end under flame resistance failure. As the syngas concentration decreased, the pressure difference between the upper and lower two ends was decreased in turn.

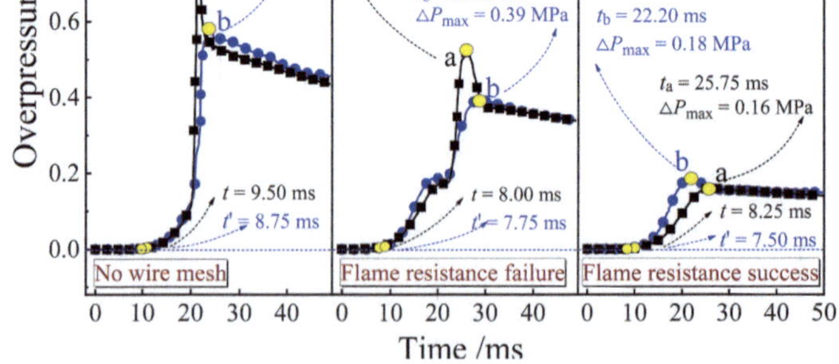

Fig. 2.6 Comparison of pressure histories under different flame resistance conditions ($c = 30\%$)

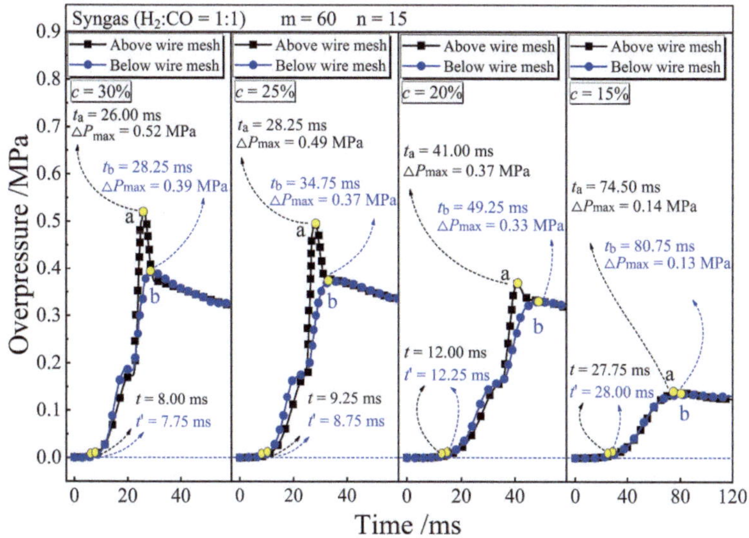

Fig. 2.7 Comparison of pressure histories under different syngas concentrations (c = 15, 20, 25 and 30%)

2.2.4 *Analysis of Explosion Pressure Rising Rate*

Figure 2.8 presents the comparison of (dP/dt) histories under different flame resistance conditions ($c = 30\%$). After adding wire mesh, (dP/dt) histories of its two ends could be evidently decreased. Although the flame resistance failure, the $(dP/dt)_{max}$ could still be decreased from 449.5 to 144.2 MPa/s^{-1} and from 329.1 to 56.6 MPa/s^{-1} respectively. In particular, it showed a more significant downward trend under flame resistance success condition and was reduced by 96.06 and 91.04%. Meanwhile, the $(dP/dt)_{max}$ of lower end was slightly higher compare with upper end. And its corresponding appearance moment also was advanced. Nevertheless, an opposite change trend was presented under the flame resistance failure condition. Especially for no wire mesh condition. It was noteworthy that the negative value of (dP/dt) history was also successively reduced with the increase of wire mesh layer number.

Figure 2.9 shows the effect of syngas concentration on (dP/dt) histories under different flame resistance condition (c = 15, 20, 25, 30%; n = 15; m = 60). As the syngas concentration decreased, the (dP/dt) history were evidently decreased in turn. It mainly reflected in the reductions of $(dP/dt)_{max}$ at the mesh wire two ends. And the $(dP/dt)_{max}$ difference between two ends was also decreased successively. In particular, the $(dP/dt)_{max}$ was reduced to 4.8 and 5.6 MPa/s^{-1} under 15% syngas concentration. And their reduction extents reached 96.67% and 90.11% respectively. Besides, a large negative value was also observed in the (dP/dt) history above wire mesh. And the negative value presented a decrease trend and its corresponding moment was also obviously delayed. This explains that the syngas explosion intensity could be

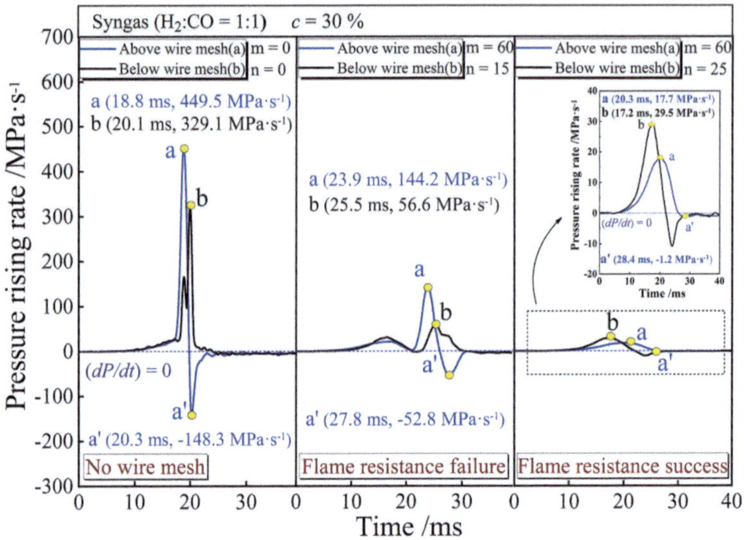

Fig. 2.8 Comparison of (dP/dt) histories under different flame resistance conditions ($c = 30\%$)

evidently decreased after adding wire mesh and the reduction degree could be also affected by syngas concentration.

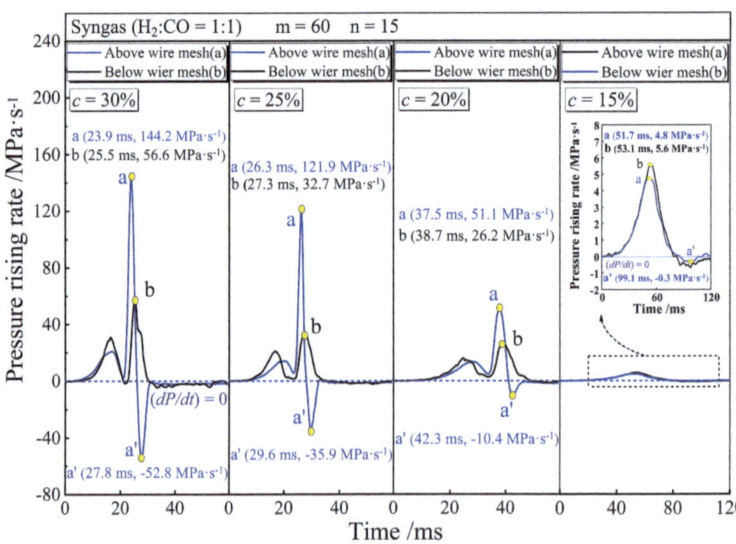

Fig. 2.9 Comparison of (dP/dt) histories under different syngas concentrations ($c = 15$, 20, 25 and 30%)

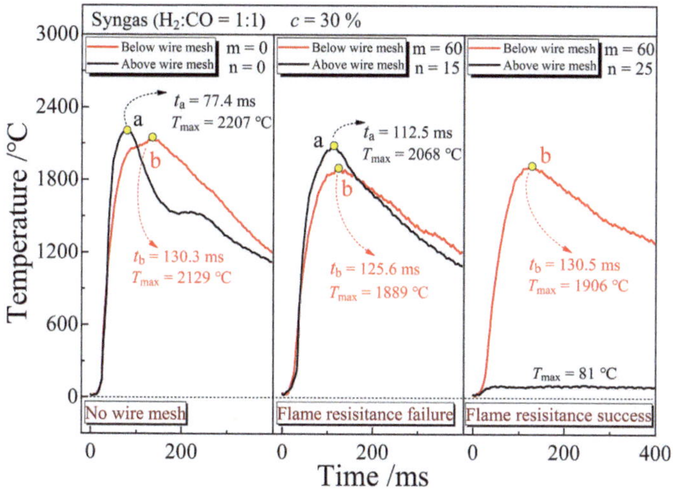

Fig. 2.10 Comparison of temperature histories under different flame resistance conditions (c = 30%)

2.2.5 *Analysis of Flame Temperature*

The comparison of flame temperature histories under different flame resistance conditions (c = 30%) is presented in Fig. 2.10. It clear that the flame temperature was significantly decreased after the wire mesh was added. As the flame resistance failed, the temperature above wire mesh was higher and the same result was also shown for the case without wire mesh. Nevertheless, the flame was effectively extinguished and failed to pass through the wire mesh as its layer number was increased. It resulted in that the temperature above wire mesh did not appear an upward change although the temperature below wire mesh showed an obvious increase. Figure 2.11 illustrates the effect of syngas concentration on flame temperature histories under the flame resistance condition (c = 15, 20, 25, 30%; n = 15; m = 60). As the syngas concentration reduced, the temperature histories were obviously decreased. T_{max} of upper and lower ends were decreased respectively from 2068 to 75 °C and from 1889 to 1546 °C, and their corresponding moments were successively prolonged. It explains that the wire mesh could evidently decrease the flame temperature of syngas explosion, and the syngas concentration could also obviously affect the flame temperature, which further affected the flame resistance effect.

2.2.6 *Analysis of Flame Resistance Mechanism*

The flame resistance mechanism is mainly attributed to the endothermic and wall effects and the flame resistance process is shown in Fig. 2.12. Through analysis, the

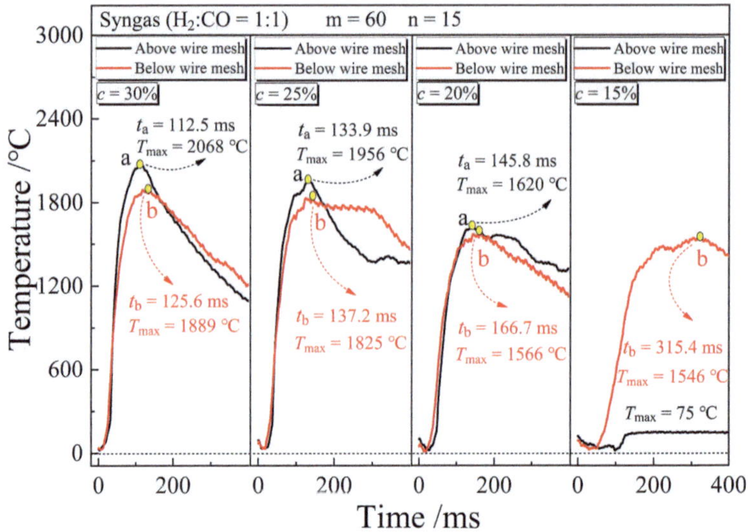

Fig. 2.11 Comparison of temperature histories under different syngas concentrations ($c = 15, 20, 25$ and 30%)

flame was divided into discrete and discontinuous micro-scale flames as it propagated inside the wire mesh, thus resulting in the obvious increase of heat dissipation area during the propagation. In addition, the heat of flame was constantly absorbed by wire mesh. The increase of contact area led to the enhancement of heat dissipation rate, further leading to the obvious decrease of temperature. Meanwhile, multistep chain reaction processes happened in the process of syngas explosion, resulting in the production of amount of high-energy radicals ($2OH = O + H_2O$, $OH + H_2 = H + H_2O$, $H + O_2 = OH + O$, $CO + OH = CO_2 + H$, $O + H_2O = OH + OH$, $CO + O_2 = CO_2 + O$, $H + O_2 = O + OH$) [7]. Nevertheless, the partial main reaction step was interrupted after the collision with wire mesh, which led to the evident decrease in the quantity and energy of high-energy free radicals. It caused an evident decrease in the collision probability of active free radical inside the wire mesh microchannel. The flame would be extinguished as the rate of active free radical destruction was greater than that of it production during the flame resistance and the temperature was lower than the ignition point of syngas.

Besides, the endothermic and wall effects could be evidently affected by the explosion intensity. Especially for the flame propagation velocity. The larger velocity would reduce the propagation time of flame inside the narrow channel, further reducing the heat dissipation effect of flame and the quenching ability of wire mesh [8]. As the syngas concentration increased, the velocity was evidently enhanced. Thus, the contact time between the flame and wire mesh and the heat dissipation was also relatively reduced. Meanwhile, the rate of free radical consumption was also relatively diminished, further leading to a rise of flame resistance failure probability. Hence, the temperature of flame and the number and energy of key free radical would

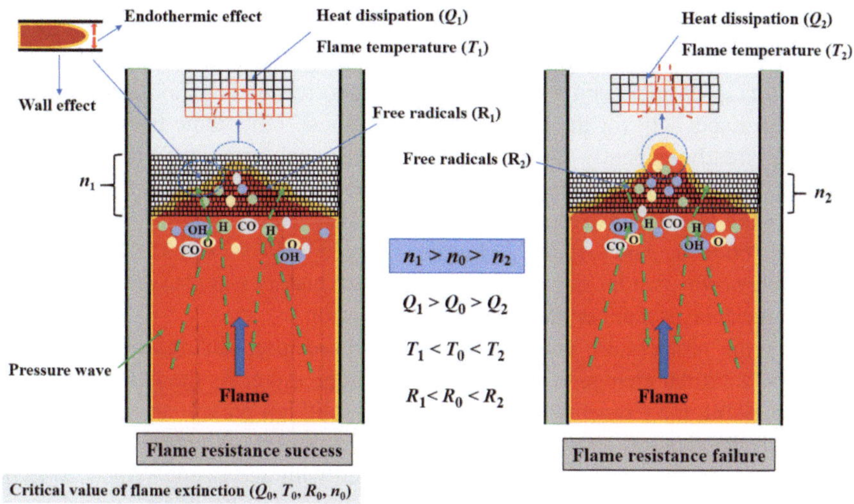

Fig. 2.12 Schematic diagram of wire mesh flame resistance process

not lower than the critical value of flame quenching, thereby causing the flame to continue to propagate. Besides, the discrete pressure wave impinged on the superimposed wire mesh and happened the reflection and scattering at different directions after entering into wire mesh, causing the greatly energy loss and intensity attenuation. And the weakening effect became more evidently as the wire mesh layer number was increased.

2.3 Conclusions

This experiment mainly focused on the influence of syngas concentration on the wire mesh flame resistance characteristics of syngas explosion by the visualization vessel. Based on the change of explosion parameter, the change of pressure rise with flame propagation above and below wire mesh under different flame resistance conditions was clarified. And the flame resistance mechanism was also analyzed. Conclusions are summarized as follows:

First, the pressure rise and flame propagation at the flame resistance system two ends showed a clear corresponding relationship during the protection. And the wire mesh could play a better inhibiting role and reduce the explosion intensity in the process of flame resistance. Flame appeared three times accelerations under flame resistance failure. Especially the acceleration phenomenon was more evident after the flame passed through wire mesh. Besides, the reflected pressure wave could pass through the wire mesh and act on the flame below wire mesh, causing the downward propagation of flame and a sharp drop of pressure.

Second, the existence of wire mesh could lead to an obvious reduction in the pressure and its rising rate histories. Under the flame resistance failure, the ΔP_{max} and $(dP/dt)_{max}$ of upper end were larger and their corresponding moments were also advanced. Especially for the case without wire mesh. The flame resistance effect could be greatly evidently affected by the syngas concentration. With decreasing syngas concentration, the time for pressure started to rise was evidently prolonged and its rising rate was also reduced. Especially, ΔP_{max} was successively reduced by 73.08 and 66.67%. And $(dP/dt)_{max}$ was also reduced by 96.06 and 91.04%. Besides, it was noteworthy that the negative value of (dP/dt) history was also greatly affected by wire mesh layer number and syngas concentration.

Third, the temperature could be greatly reduced after adding wire mesh and presented a similar change trend to the explosion pressure. As the flame resistance was successful, the temperature above wire mesh did not present an increase trend and reach the ignition point of syngas. T_{max} was only 81 °C. Besides, the flame temperatures of two ends were obviously reduced with decreasing syngas concentration and affected by the layer number of wire mesh. It clear that the flame resistance mechanism was mainly attributed to the combination of the endothermic and wall effects.

For effective prevention of syngas explosion with the wire mesh flame resistance system, three suggestions are put forward: (1) To realize the effective flame resistance, the material that can enhance the quenching effect of wire mesh on the flame should be preferentially adopted. (2) For the flame resistance failure, the explosion intensity of flame resistance system downstream was higher compare with its upstream. Hence, the double inhibition methods should be considered to achieve effective explosion protection. (3) The flame resistance effect was related to its explosion intensity. As the metal wire mesh was selected to prevent flame, the concentration of premixed gas should be reduced as much as possible during the industrial production.

References

1. Cao XY, Wang Z, Wang Y, Lu YW, Sun SC, Xu JJ. Experimental research on the flame resistance characteristics of wire mesh for syngas explosion. Process Saf Environ Protect. 2023;175:34–47.
2. Yang XF, Yu MG, Zheng K, Luan PP, Han SX. On the propagation dynamics of lean H_2/CO/air premixed flame. Int J Hydrog Energy. 2020;45(11):7210–22.
3. Xiao HH, Wang QS, Shen XB, An WG, Duan QL, Sun JH. An experimental study of premixed hydrogen/air flame propagation in a partially open duct. Int J Hydrog Energy. 2014;39(11):6233–41.
4. Yu ZJ, Chen SS, Gu Y, Wen H, Li RK, Fan SX. Premixed flame propagation of methane/carbon monoxide/air in a closed tube with an obstacle. Combust Sci Technol. 2020;193(15):2724–40.
5. Jin KQ, Duan QL, Chen JY, Liew KM, Gong L, Sun JH. Experimental study on the influence of multi-layer wire mesh on dynamics of premixed hydrogen-air flame propagation in a closed duct. Int J Hydrog Energy. 2017;42(21):14809–20.
6. Cheng FM, Lu JJ, Li TT, Luo ZM. Study on the influence of obstacle on the flame behaviors of methane/air in a semi-confined pipe with wire mesh installed. J Loss Prev Process Ind. 2022;78:104826.

7. Li WY, Zou C, Yao H, Lin QJ, Fu R, Luo JH. An optimized kinetic model for H_2/CO combustion in CO_2 diluent at elevated pressures. Combust Flame. 2022;241:112093.
8. Yue JP, Wei L, Liu HG, Guo SS. A novel detonation arrester containing a large disk with long triangular slits: design and numerical simulation. Process Saf Prog. 2020;10:1–16.

Chapter 3
Effect of Wire Mesh Structure Parameter on the Flame Propagation Characteristics of Syngas Explosion

3.1 Experimental Apparatus and Procedures

Based on the visual experimental apparatus built in the previous work [1], the experiment on the effect of wire mesh structure on syngas explosion inhibition was conducted. The detailed schematic diagram is shown in chapter one. The main difference from the previous apparatus was that a wire mesh flame resistance system was installed inside the explosion vessel (910 mm × 150 mm × 150 mm) and fixed by four support rods ($d = 5$ mm). The wire mesh flame resistance system was composed of the multi-layer wire mesh and two support plates ($\delta = 3$ mm). The wire mesh with 100 mm × 100 mm effective flame resistance area was placed inside the two support plates and was compressed by the bolt connection method. The wire mesh flame resistance system was installed at a height of 650 mm (the middle of two pressure transmitters and thermocouples) distance from the bottom end of the vessel. Through previous researches, the height could ensure the acceleration propagation and full development of syngas explosion flame.

In this experiment, the effect of wire mesh structure on the flame propagation characteristics and intensity of syngas explosion was studied deeply based on the changes of explosion parameters at the upper and lower two ends of wire mesh. The dynamic evolution process of flame under different wire mesh structure parameters was recorded by a high-speed camera. Considering the propagation velocity of syngas explosion flame, the acquisition frequency was set to 4000 fps. Two high-frequency pressure transmitters with 50 kHz response frequency and 0.25% acquisition accuracy and two S-type high-frequency thermocouples with 3000 °C range were respectively installed in the upper and lower wall surfaces of wire mesh flame resistance system [the distance (L) = 50 mm] to acquire the pressure and temperature data. An eight-channel high frequency data acquisition card with 200 kHz was adopted to realize the program control and data acquisition.

Four kinds of syngas concentrations (CO/H_2 = 1:1; c = 15, 20, 25 and 30%) was selected during the experiment and prepared inside the explosion vessel by

Z. Wang and X. Cao, *Flame Resistance Characteristics and Its Influence Mechanism*,
https://doi.org/10.1007/978-981-95-6119-3_3

the Dalton's law of partial pressure. And four kinds of wire mesh numbers (m = 0, 40, 60 and 80) and five kinds of wire mesh layers (n = 0, 15, 20, 25 and 30) were adopted respectively during the explosion inhibition. The initial pressure was 0.1 MPa at the ignition moment. After ignition, the premixed gas of 110 mm height distance from the bottom end of the vessel was ignited (Ignition energy = 8 kV) and propagated upward. The flame would contact with the wire mesh during its propagation. The flame structure, explosion intensity and flame temperature could be obviously changed after passing through the flame resistance system. To ensure the reliability and accuracy of the experiment, each group of experiments was repeated 3–5 times under the same working conditions.

3.2 Results and Discussions

3.2.1 Flame Resistance Characteristics of Wire Mesh

Figure 3.1 illustrates the corresponding relationship between the flame propagation and pressure rise of syngas explosion at the upper and lower two ends of wire mesh under flame resistance success and failure conditions (c = 25%; n = 15; m = 40 and 80). As can be seen from Fig. 3.1a, the flame passed through the wire mesh, which meant the flame resistance failure (c = 25%; n = 15; m = 40). However, an obvious corresponding relationship between the flame propagation and pressure rise was showed. The flame developed as the light blue "spherical" structure after the premixed gas was ignited. The initial explosion flame was weak and far away from the pressure transmitter, resulting in the slow flame propagation velocity and no obvious pressure rise. After that, the flame surface area was increased and its propagation was accelerated due to the sidewall restraint. As the flame propagation accelerated, the pressure history of wire mesh lower end began to rise ($t = 7.50$ ms). However, the moment when the pressure history of wire mesh upper end began to rise was slight delayed ($t = 8.25$ ms). With the change of flame from the "sphere" to "finger" structure, its propagation velocity was increased obviously. The pressure also appeared an obvious increase process. And the increase extent of lower end was more significant in the early stage.

As the flame front curvature radius increased, the flame propagation was slowed down [2]. The flame appeared the second acceleration process as it approached to the wire mesh. And it passed through the wire mesh and ignited the premixed gas of upper end after the flame stagnated for 1 ms inside the wire mesh. After that, the flame was significant accelerated and the velocity peak reached 225.8 m/s. Due to the flame resistance failure, the pressure histories at the upper and lower ends of wire mesh also presented a more significant increase process. In particular, the increase extent of upper end was more obviously (dP/dt = 187.7 MPa/s). And the ΔP_{max} was higher than that of lower end. However, the reverse pressure wave would result in an obvious pressure drop and a large negative value in pressure rising rate history

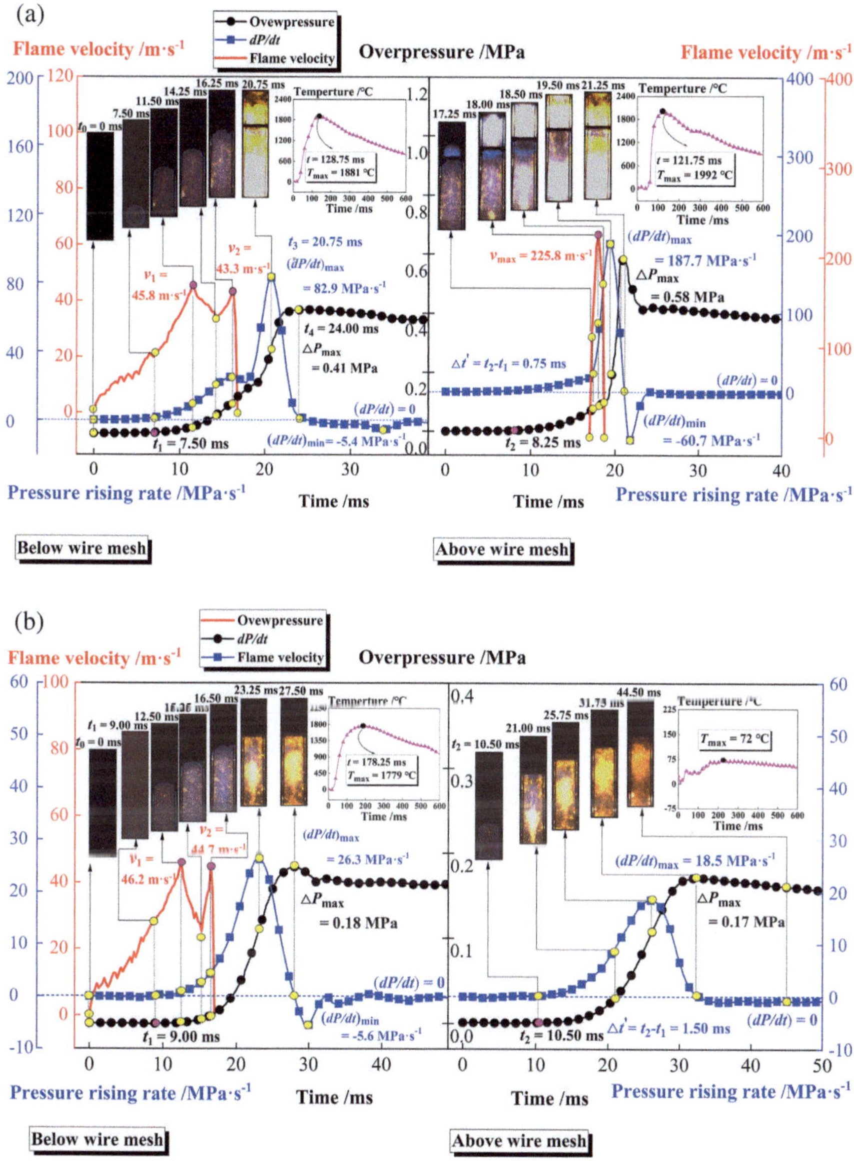

Fig. 3.1 **a** Corresponding relationship between flame propagation and pressure rise of syngas explosion at the upper and lower two ends of wire mesh under flame resistance failure condition ($c = 25\%$; n = 15; m = 40). **b** Corresponding relationship between flame propagation and pressure rise of syngas explosion at the upper and lower two ends of wire mesh under flame resistance success condition ($c = 25\%$; n = 15; m = 80)

($(dP/dt)_{min} = -60.7$ MPa/s). Besides, the T_{max} was slightly higher compared with the lower end. After that, the pressure history was decreased obviously as the flame extinguished [3].

Figure 3.1b shows the corresponding relationship between the flame propagation and pressure rise of syngas explosion at the upper and lower two ends of wire mesh under flame resistance success condition ($c = 25\%$; n = 15; m = 80). It can be seen that an obvious corresponding relationship was also presented during the explosion inhibition. The acceleration propagation of flame could lead to the sharp increase of pressure history. The flame also presented the same change process from the "spherical" to "finger" structure. However, the moment when the pressure histories of upper and lower two ends began to rise was slight delayed compared with the flame resistance failure condition ($\Delta t = 1.5$ and 2.25 ms). And the velocity history also appeared a decrease process and the decrease extent was more significant due to the increase of wire mesh number. The flame appeared the second acceleration process as it approached to the wire mesh, resulting in a rapid increase of pressure rising rate history. A stagnation phenomenon was also presented inside the wire mesh. However, the flame did not pass through it to ignite the premixed gas of upper end. Besides, the explosion intensity appeared an obvious decrease as flame resistance was successful. For instance, the ΔP_{max} of upper and lower ends were decreased from 0.41 MPa and 0.58 MPa to 0.18 MPa and 0.17 MPa respectively. And the $(dP/dt)_{max}$ were also decreased from 82.9 MPa/s and 187.7 MPa/s to 26.3 MPa/s and 18.5 MPa/s respectively. However, the drop rate of pressure history at the upper end of wire mesh was obviously reduced compared with the flame resistance failure. And the decrease extent of flame temperature was more obviously. After that, the pressure history was sharply decreased with the acceleration extinction of explosion flame.

3.2.2 *Analysis of Flame Propagation Process*

Figure 3.2 presents the comparison of flame propagation process under different mesh number conditions ($c = 25\%$; n = 15; m = 40, 60 and 80). The syngas explosion flame could be effectively inhibited with increasing mesh number. And the flame all experienced the acceleration process and a change from "spherical" to "finger" structure at the initial stage of flame propagation. Combined flame propagation velocity history (in Fig. 3.3), the flame experienced three times acceleration processes as the flame resistance failed (I, II and III stages). Namely, the self-accelerating propagation, mesh-accelerating propagation and turbulence-accelerating propagation processes. As the flame resistance was successful, the flame only experienced twice acceleration processes (I and II stages).

Meanwhile, the flame propagation velocity presented a change trend of the decrease firstly and then increase with the increase of mesh number. It because that the acceleration propagation of flame was mainly affected by the flame self-acceleration and the flame front disturbance due to the action of reflected pressure wave generated from the blocking of wire mesh [4]. Besides, the flame appeared

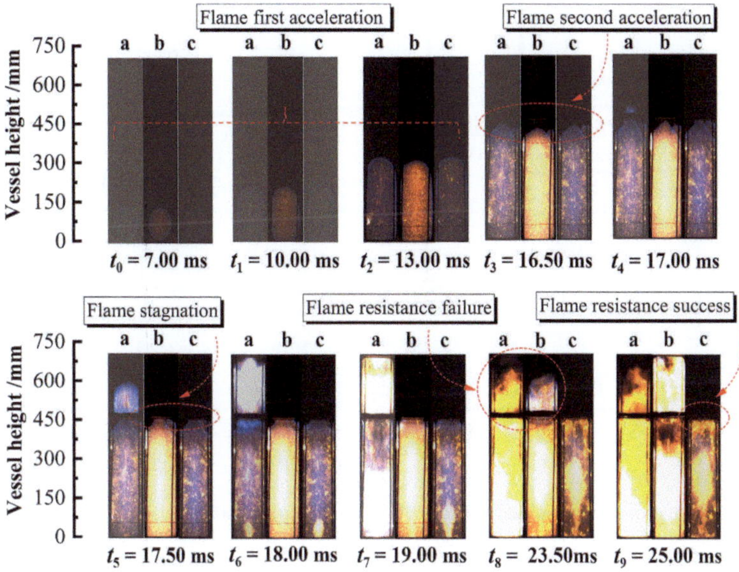

Fig. 3.2 The comparison of flame propagation process under different mesh number conditions ($c = 25\%$; n = 15; m = 40, 60 and 80)

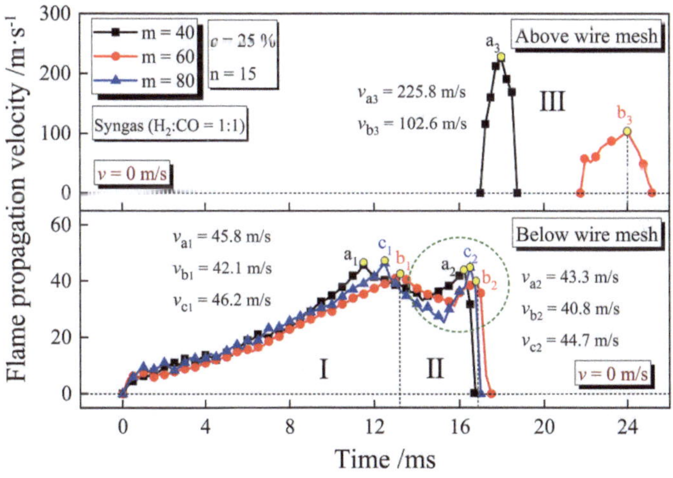

Fig. 3.3 Flame propagation velocity history of wire mesh upper and lower two ends with increasing mesh number ($c = 25\%$; n = 15; m = 40, 60 and 80)

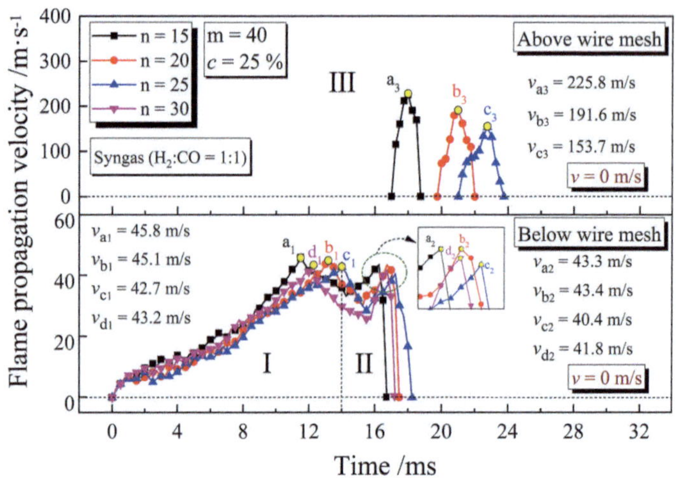

Fig. 3.4 Flame propagation velocity history of wire mesh upper and lower two ends with increasing mesh layer ($c = 25\%$; m = 40; n = 15, 20, 25 and 30)

an obvious acceleration after passing through the wire mesh. However, the velocity history at the upper end of wire mesh was decreased significantly with increasing mesh number.

Figure 3.4 shows the comparison of flame propagation velocity history at the upper and lower ends of wire mesh under different mesh layer conditions ($c = 25\%$; m = 40; n = 15, 20, 25 and 30). The flame propagation also experienced three times acceleration processes and presented a change trend of decrease with the increase of mesh number as flame resistance failed. Especially, the decrease in the flame propagation velocity of wire mesh upper end was more significantly (from 225.8 to 153.7 m/s). It indicates that the structure parameter of wire mesh (the mesh number and layer) had a more significant acceleration effect on the flame propagation velocity of its rear end. As n = 30, the syngas explosion flame could be effectively inhibited and the flame occurred the stagnation inside the wire mesh. However, the disturbance of flame front was significantly enhanced due to the enhancement of reflection pressure wave with increasing mesh layer, resulting in a slight rise trend of flame propagation velocity.

3.2.3 Analysis of Explosion Pressure

The inhibition effect of wire mesh on syngas explosion was mainly reflected in the reduction extent of explosion intensity at the two ends of wire mesh. Figure 3.5 presents the changes of explosion pressure histories at the upper and lower ends of wire mesh with increasing mesh layer ($c = 25\%$; m = 40; n = 0, 15, 20, 25 and 30). It can be seen that the pressure at the upper end of wire mesh was obviously greater

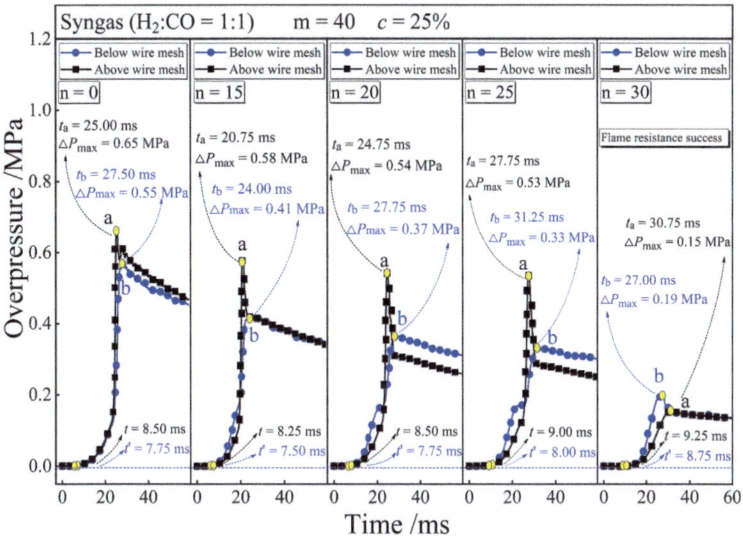

Fig. 3.5 The changes of pressure histories at the upper and lower two ends of wire mesh with increasing mesh layer ($c = 25\%$; m = 40; n = 0, 15, 20, 25 and 30)

than that of the lower end as the flame resistance failed. However, an opposite result was presented as the flame resistance was successful. As the wire mesh existed, the explosion pressure was obviously decreased (from 0.65 and 0.55 to 0.58 and 0.41 MPa). And ΔP_{max} was continuously reduced from 0.58 to 0.53 MPa and from 0.41 to 0.33 MPa with increasing mesh number. Especially, the pressure drop was more significant as the flame resistance was successful ($\Delta P_{max} = 0.15$ and 0.19 MPa). ΔP_{max} of upper and lower ends were reduced by 76.9% and 65.5% respectively. And the moment when the pressure history of upper end began to rise was slightly delayed than that of lower end.

Figure 3.6 illustrates the changes of explosion pressure histories at the upper and lower ends of wire mesh with increasing mesh number ($c = 25\%$; n = 15; m = 0, 40, 60 and 80). The pressure at the upper end of wire mesh was also greater than that of the lower end as the flame resistance failed. And the pressure was gradually decreased as wire mesh number increased (from 0.65 and 0.55 to 0.49 and 0.37 MPa). In particular, the pressure was significantly reduced as the flame resistance was successful. And the reduction extent reached 73.8% and 67.3% respectively. Meanwhile, the pressure at the lower end of wire mesh was slight greater than that of the upper end. It indicates that the increase of wire mesh number could obviously reduce the explosion pressure at its the front and rear ends. And the decrease of explosion intensity was most significantly as flame resistance was successful ($\Delta P_{max} = 0.18$ and 0.17 MPa). Besides, the moment when the pressure histories of wire mesh two ends began to rise and the appearance moment of ΔP_{max} were also slightly delayed with increasing wire mesh number. The similar change trend was also presented under the four syngas concentrations with increasing wire mesh number and layer (see in Table 3.1).

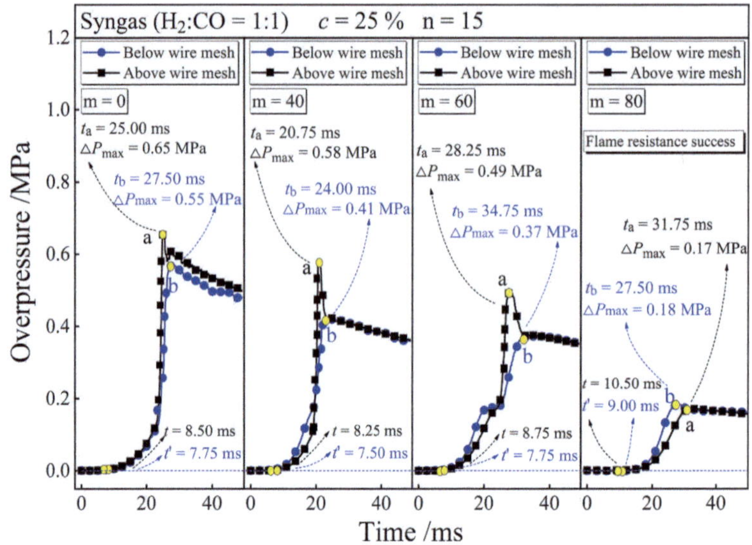

Fig. 3.6 The changes of pressure histories at the upper and lower two ends of wire mesh with increasing mesh number (c = 25%; n = 15; m = 0, 40, 60 and 80)

Table 3.1 ΔP_{max} under different syngas concentrations with increasing wire mesh number and layer

Wire mesh structure parameter			Above wire mesh			Below wire mesh		
			m = 40	60	80	m = 40	60	80
c=30%	n = 0	$\triangle P_{max}$	0.81	0.81	0.81	0.58	0.58	0.58
	n = 15	$\triangle P_{max}$	0.66	0.52	0.51	0.44	0.39	0.33
	n = 20	$\triangle P_{max}$	0.63	0.51	0.16	0.38	0.39	0.19
	n = 25	$\triangle P_{max}$	0.62	0.16	0.15	0.35	0.18	0.20
	n = 30	$\triangle P_{max}$	0.15	0.15	0.15	0.20	0.19	0.21
c=25%	n = 0	$\triangle P_{max}$	0.65	0.65	0.65	0.55	0.55	0.55
	n = 15	$\triangle P_{max}$	0.58	0.49	0.17	0.41	0.37	0.18
	n = 20	$\triangle P_{max}$	0.54	0.46	0.16	0.37	0.35	0.19
	n = 25	$\triangle P_{max}$	0.53	0.16	0.16	0.33	0.18	0.19
	n = 30	$\triangle P_{max}$	0.15	0.16	0.15	0.19	0.19	0.20
c=20%	n = 0	$\triangle P_{max}$	0.48	0.48	0.48	0.47	0.47	0.47
	n = 15	$\triangle P_{max}$	0.55	0.37	0.16	0.35	0.33	0.16
	n = 20	$\triangle P_{max}$	0.52	0.47	0.16	0.33	0.32	0.16
	n = 25	$\triangle P_{max}$	0.50	0.15	0.15	0.30	0.15	0.16
	n = 30	$\triangle P_{max}$	0.15	0.15	0.15	0.17	0.16	0.18
c=15%	n = 0	$\triangle P_{max}$	0.37	0.37	0.37	0.37	0.37	0.37
	n = 15	$\triangle P_{max}$	0.31	0.13	0.14	0.30	0.14	0.14
	n = 20	$\triangle P_{max}$	0.32	0.14	0.14	0.29	0.14	0.14
	n = 25	$\triangle P_{max}$	0.31	0.14	0.13	0.26	0.13	0.12
	n = 30	$\triangle P_{max}$	0.13	0.14	0.14	0.13	0.14	0.13

3.2.4 Analysis of Explosion Pressure Rising Rate

The reduction of explosion intensity could be directly reflected by the change of pressure rising rate. Figure 3.7 shows the changes of pressure rising rate histories at the upper and lower ends of wire mesh with increasing mesh layer ($c = 25\%$; m = 40; n = 0, 15, 20, 25 and 30). The change of pressure rising rate history was similar to that of the explosion pressure history. The history at wire mesh lower end appeared two peak values as the flame resistance failed and the change became more significant with increasing wire mesh layer. However, the history at wire mesh upper end only presented a peak value. With increasing mesh layer, these peak values all were decreased obviously. In particular, the pressure rising rate history of wire mesh upper end appeared a negative value and constantly reduced with increasing mesh layer. It indicates that the explosion intensity was sharply decreased as the wire mesh existed and the effect was enhanced ceaselessly with the increase of mesh layer. As n = 30, the reduction of explosion intensity was most significant as the flame resistance was successful. Figure 3.8 presents the changes of pressure rising rate histories at the upper and lower ends of wire mesh with increasing mesh number ($c = 25\%$; n = 15; m = 0, 40, 60 and 80). It can be seen that the change was similar to that of the pressure rising rate history with increasing mesh layer. The pressure rising rate history of wire mesh upper end also appeared a large negative value. However, the decrease extent was gradually reduced with increasing mesh number. As m = 80, the flame could be effectively inhibited and the explosion reaction rate could be also obviously decreased. The same change trend was also presented under the four syngas concentrations with increasing wire mesh number and layer (see in Table 3.2).

3.2.5 Analysis of Flame Temperature

Figure 3.9 presents the changes of flame temperature histories at the upper and lower ends of wire mesh with increasing mesh layer ($c = 25\%$; m = 40; n = 0, 15, 20, 25 and 30). As the wire mesh existed, the flame temperature was obviously reduced (from 2174 to 1992 °C). The flame temperature appeared a slight decrease trend as flame resistance failed and the temperature of wire mesh upper end was higher than that of lower end. Compared with the flame resistance failure, the temperature change of wire mesh upper end was more significant as the flame resistance was successful. The temperature history did not rise obviously. T_{max} only was 90 °C and did not reached the ignition temperature of syngas. Hence, the premixed gas of wire mesh upper end could not be ignited.

Figure 3.10 shows the changes of flame temperature histories at the upper and lower ends of wire mesh with increasing mesh number ($c = 25\%$; n = 15; m = 0, 40, 60 and 80). The similar trend was presented and the flame temperature was also decreased obviously as the wire mesh existed. With the increase of mesh number, the flame temperature could be obviously decreased. Especially, the flame could be

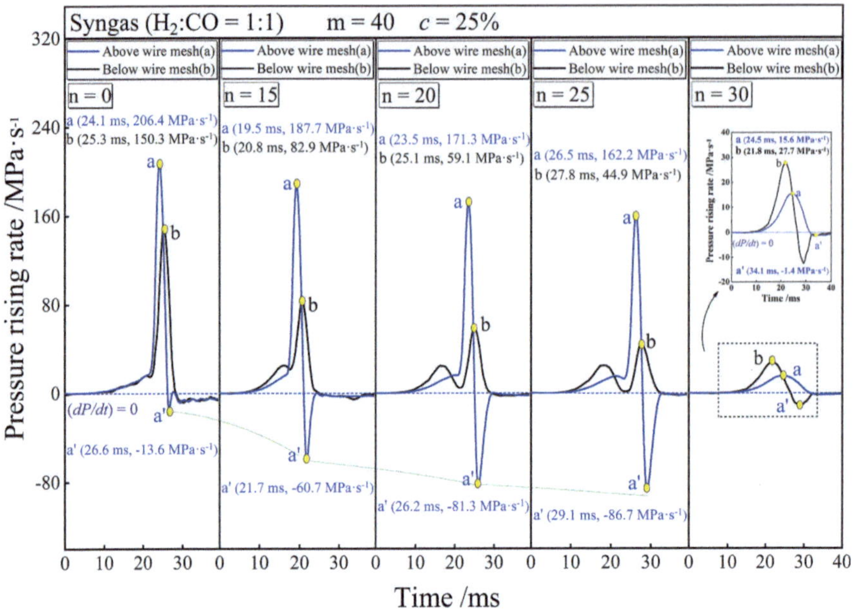

Fig. 3.7 The changes of pressure rising rate histories at the upper and lower ends of wire mesh with increasing mesh layer (c = 25%; m = 40; n = 0, 15, 20, 25 and 30)

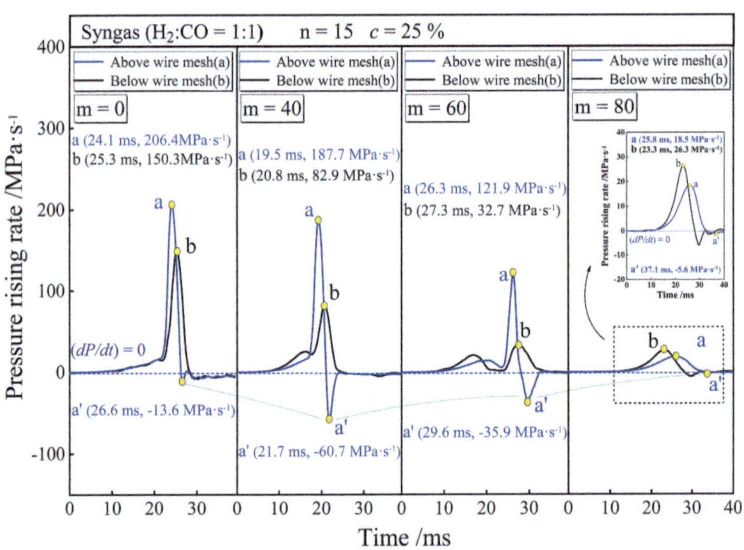

Fig. 3.8 The changes of pressure rising rate histories at the upper and lower ends of wire mesh with increasing mesh number (c = 25%; n = 15; m = 0, 40, 60 and 80)

Table 3.2 $(dP/dt)_{max}$ under different syngas concentrations with increasing wire mesh parameter

Wire mesh structure parameter			Above wire mesh			Below wire mesh		
			m = 40	60	80	m = 40	60	80
c=30%	n = 0	$(dP/dt)_{max}$	449.5	449.5	449.5	329.1	329.1	329.1
	n = 15	$(dP/dt)_{max}$	217.2	144.2	141.6	85.9	56.6	38.4
	n = 20	$(dP/dt)_{max}$	192.1	147.3	17.5	55.4	48.9	29.1
	n = 25	$(dP/dt)_{max}$	173.9	17.7	16.6	41.6	29.5	28.0
	n = 30	$(dP/dt)_{max}$	15.8	16.6	15.7	29.4	28.6	29.2
c=25%	n = 0	$(dP/dt)_{max}$	206.4	206.4	206.4	150.3	150.3	150.3
	n = 15	$(dP/dt)_{max}$	187.7	121.9	18.5	82.9	32.7	26.3
	n = 20	$(dP/dt)_{max}$	171.3	113.6	16.7	59.1	30.1	26.7
	n = 25	$(dP/dt)_{max}$	162.2	16.4	15.8	44.9	24.1	26.3
	n = 30	$(dP/dt)_{max}$	15.6	16.5	15.1	27.7	26.8	26.6
c=20%	n = 0	$(dP/dt)_{max}$	90.1	90.1	90.1	88.3	88.3	88.3
	n = 15	$(dP/dt)_{max}$	150.3	51.1	12.9	47.2	26.2	15.4
	n = 20	$(dP/dt)_{max}$	136.4	102.6	12.9	40.4	37.1	16.7
	n = 25	$(dP/dt)_{max}$	125.7	12.5	11.9	30.7	15.4	15.8
	n = 30	$(dP/dt)_{max}$	12.9	12.9	12.8	19.3	17.2	19.2
c=15%	n = 0	$(dP/dt)_{max}$	24.4	24.4	24.4	24.1	24.1	24.1
	n = 15	$(dP/dt)_{max}$	54.8	5.6	6.1	33.9	4.8	6.5
	n = 20	$(dP/dt)_{max}$	51.7	5.8	7.1	26.3	6.1	7.5
	n = 25	$(dP/dt)_{max}$	47.4	6.1	5.6	18.3	6.3	6.1
	n = 30	$(dP/dt)_{max}$	5.9	5.7	6.3	7.9	6.3	6.6

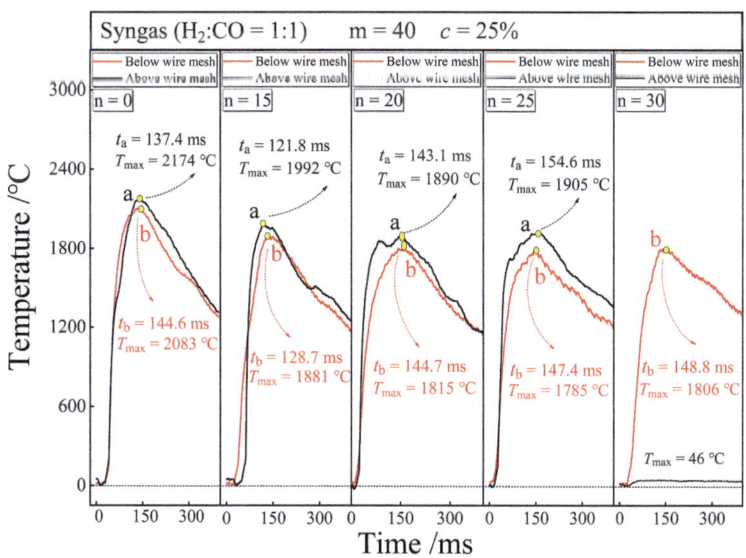

Fig. 3.9 The changes of flame temperature histories at the upper and lower ends of wire mesh with increasing mesh layer (c = 25%; m = 40; n = 0, 15, 20, 25 and 30)

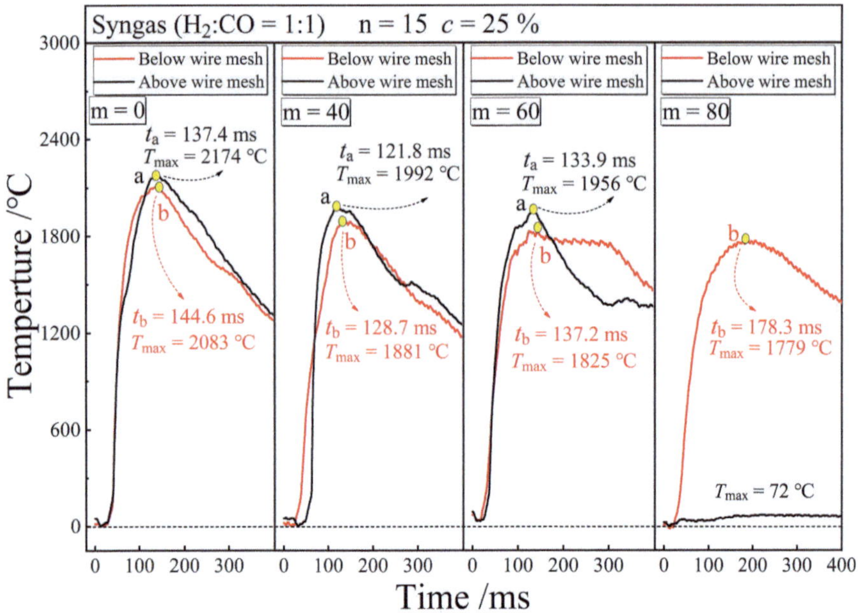

Fig. 3.10 The changes of flame temperature histories at the upper and lower ends of wire mesh with increasing mesh number ($c = 25\%$; n = 15; m = 0, 40, 60 and 80)

completely inhibited as m = 80. And the temperature history at the upper end of wire mesh was greatly decreased and T_{max} was only 70 °C. Hence, the premixed gas could not be ignited.

3.2.6 Analysis of Flame Resistance Mechanism

The inhibition of wire mesh on syngas explosion was mainly attributed to the combination of the endothermic action and wall effect. Figure 3.11 presents the schematic diagram of flame resistance mechanism by wire mesh. The smooth flame would be dispersed into the large number of small flame unit after entering the metal mesh narrow channel. As an excellent heat conduction medium, the wire mesh could quickly absorb the heat of flame and reduce the intensity of shock wave. And the dispersion extent of flame could be enhanced with the increases of mesh number and layer, resulting the obvious increases in the contact area between the flame and wire mesh and their heat exchange rate. Meanwhile, the flame propagation path in the narrow channel was also increased obviously as the wire mesh thickness was increased, further resulting in the obvious weaken of flame propagation capacity.

Besides, the syngas explosion was accompanied by the chain reaction process of a large number of active free radicals. Amount of high-energy radicals would be

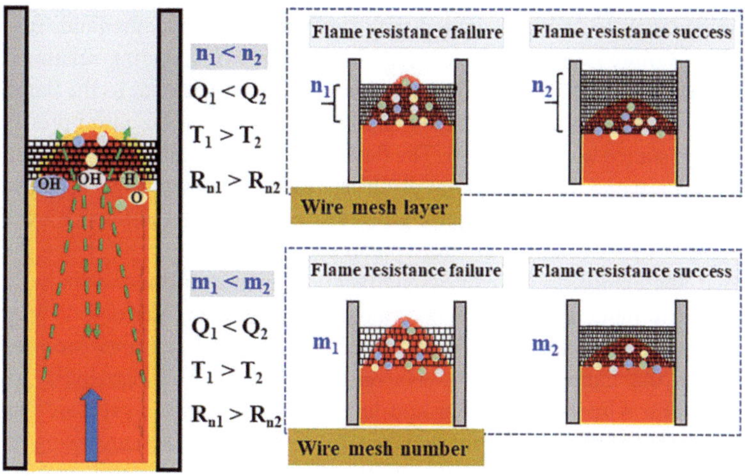

Fig. 3.11 Schematic diagram of flame resistance mechanism by wire mesh

generated to promote the chain reaction during the syngas explosion ($CO + O_2 = CO_2 + O$, $O + H_2O = OH + OH$, $CO + OH = CO_2 + H$, $H + O_2 = OH + O$, $H + O_2 = O + OH$, $OH + H_2 = H + H_2O$, $2OH = O + H_2O$). However, the key reaction process was interrupted and the energy and quantity of active free radicals were obviously reduced after contacting with wire mesh. It would greatly promote the reduction in collision probability of active free radicals and the loss of energy [5, 6]. Hence, the flame propagation capacity was weakened and the velocity was reduced obviously. The propagation time of flame inside the wire mesh and the heat dissipation were relatively increased with the increases of wire mesh number and layer. And the number of free radical involved in the chain reaction was also relatively reduced, further resulting in an increase of flame resistance success probability. In particular, the temperature of flame front and the number of chain reaction free radical would be lower than the critical value of flame extinction, thus causing the flame extinction.

3.3 Conclusions

The changes of the flame propagation characteristics and explosion parameter and their corresponding relationship at two ends of wire mesh were studied by the visual explosion inhibition experiment under different the wire mesh number and layer conditions. And the influence law and mechanism of wire mesh structure on syngas explosion were analyzed by the comparison under the flame resistance success and failure conditions. Conclusions are as follows.

First, the propagation of syngas explosion flame could be effectively inhibited by the wire mesh. And the inhibition effect was related to the wire mesh structure. The flame propagation and pressure rise of wire mesh two ends under the flame resistance success and failure conditions presented an obvious corresponding relationship. In addition, the flame experienced three times acceleration processes as the flame resistance failed. However, the flame only experienced twice acceleration processes as the flame resistance was successful. The flame propagation was affected by the flame self-acceleration and the flame front disturbance due to the action of reflected pressure wave generated from the blocking of wire mesh. And the influence extent was also related to the wire mesh structure parameter.

Second, the explosion intensity could be effectively weakened as the wire mesh existed. As the flame resistance failed, the pressure at the upper end of wire mesh was obviously greater than that of the lower end. However, an opposite result was presented as the flame resistance was successful. ΔP_{max} was obviously reduced with increasing wire mesh number and layer. In particular, the pressure drop was more significant as the flame resistance was successful. The pressure rising rate also presented a change law similar to that of ΔP_{max} and was more significantly affected by wire mesh structure. Besides, the flame temperature was obviously reduced as the wire mesh existed. The flame temperature appeared a slight decrease trend as flame resistance failed and the temperature of wire mesh upper end was also higher than that of lower end. Compared with the flame resistance failure, the temperature change of wire mesh upper end was more significant as the flame resistance was successful and did not reach the ignition temperature of syngas.

For the practical application of wire mesh on effective inhibition of syngas explosion during the safety protection of chemical process equipment, some suggestions are given according to present experiments: (1) To better inhibit flame propagation and decrease pressure drop under normal working condition, the optimal wire mesh structure parameter should be selected as far as possible. (2) The material with a high thermal conductivity should be preferentially selected as the wire mesh flame retardant materials. (3) Chemical reagents that can improve the wall effect should be sprayed on the wire mesh surface to the enhance the inhibition effect.

References

1. Cao XY, Wang ZR, Wang Y, Wang ZR, Zhou Y. Effect of wire mesh structure parameter on the flame propagation characteristics of syngas explosion. Fuel. 2023;334:126658.
2. Yang XF, Yu MG, Zheng K, Luan PP, Han SX. On the propagation dynamics of lean H_2/CO/air premixed flame. Int J Hydrogen Energy. 2020;45(11):7210–22.
3. Wang SX, Fan AW. Effect of CO-to-H_2 ratio on syngas flame behaviors in a micro flow reactor with controlled wall temperature gradient. Fuel. 2023;331(2):125959.
4. Li Y, Zhao Q, Liu L, Chen X, Huang C, Yuan B. Investigation on the flame and explosion suppression of hydrogen/air mixtures by porous copper foams in the pipe with large aspect ratio. J Loss Prev Process Ind. 2022;76:104744.
5. Lin C, Cao X, Wang Z, Wei J, Xu J. Research on quenching performance and multi-factor influence law of hydrogen crimped-ribbon flame arrester using response surface methodology. Fuel. 2022;326:124911.

6. Davis SG, Joshi AV, Wang H, Egolfopoulos F. An optimized kinetic model of H_2/CO combustion. Proc Combust Inst. 2005;30(1):1283–92.

Chapter 4
Wire-Mesh Inhibition of Jet Fire Induced by Explosion Venting

4.1 Experimental Apparatus and Procedures

As shown in Fig. 4.1, the experimental system consists of an explosion venting system, a data acquisition system, and a gas distribution system [1].

4.1.1 Experimental Venting System

In this paper, the material comprising the explosion venting vessel device is Q 345 R steel, which has good anti-explosion performance. The vessel has a length of 424 cm, an inner diameter of 424 cm, a length-diameter ratio of 1:1, and a volume of 55 L. There are 4 M 20 × 1.5 threaded interfaces on the vessel wall, which are used to install a pressure sensor, a flame sensor, inlet and outlet air valves and an ignition rod. The outlets at both ends of the vessel are two reinforcing pipes that can be connected with other pipes by flanges or blocked by blind plates. The length of the reinforcing pipes is 125 mm; these pipes are arranged symmetrically, and each reinforcing pipe has two thread interfaces of M 20 × 1.5. The type of igniter is KTD-A, which can adjust the ignition energy and output energy from 5 to 20 J.

In this paper, 40, 60, 80 and 100 mesh wire mesh were selected for experimental study. The material of the wire mesh we choose is 304 stainless steel, and its melting point is about 1400 °C. The specific heat capacity of 304 stainless steel at 0–100 °C is 0.50 kJ kg^{-1} K^{-1}. The thermal conductivity of 304 stainless steel at 100 and 500 °C is 16.3 W m^{-1} K^{-1} and 21.5 W m^{-1} K^{-1} respectively. Methane explosion is a process that is completed in an instant, and there is a grid in the middle of the wire mesh, so the wire mesh can withstand the high temperature and high pressure brought by the methane explosion. We will promptly replace it after every experiment. Different types of wire mesh parameters are shown in Table 4.1.

Z. Wang and X. Cao, *Flame Resistance Characteristics and Its Influence Mechanism*,
https://doi.org/10.1007/978-981-95-6119-3_4

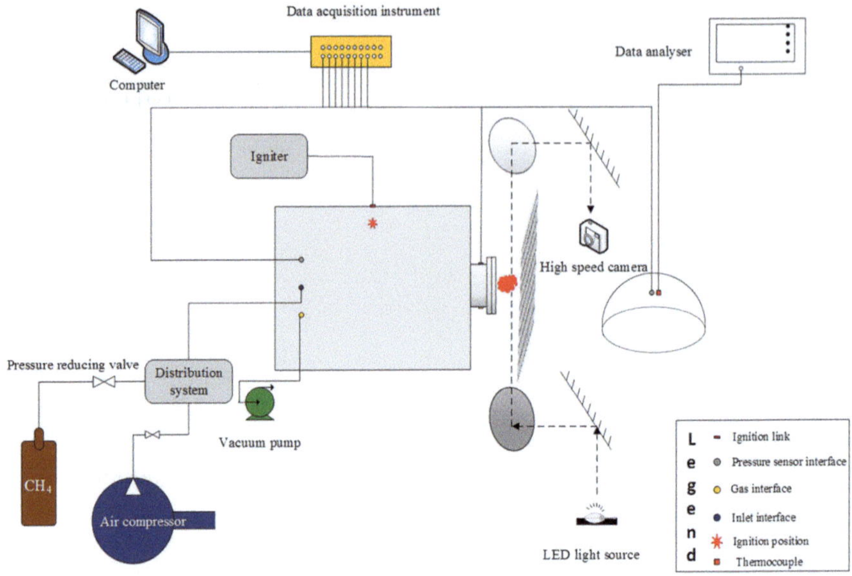

Fig. 4.1 Schematic diagram of wire-mesh suppression device for jet fire induced by explosion venting

Table 4.1 Different types of wire mesh parameters

Number of wire mesh	Number of holes per centimeter	The aperture size (mm)	Wire diameter size (mm)	Metal surface integral rate	Volume fraction of metal
40	15.7	0.439	0.194	0.4856	0.4776
60	23.6	0.302	0.154	0.4114	0.5736
80	31.5	0.215	0.104	0.4605	0.5044
100	39.4	0.172	0.081	0.4622	0.5007

4.1.2 Data Acquisition System

In this paper, the flame temperature acquisition uses a platinum rhodium alloy wire made of a thermocouple and an HIOKI8860-50 high-speed data acquisition instrument. The sensor used for pressure collection is a KISTLER 211B5 pressure transmitter with a measuring range of 0–0.7 MPa. A SIRIUS type data acquisition instrument from DEWESoft is used for pressure data acquisition, which has 16 input/output channels, high measurement accuracy and a sampling rate of 200 kS/s/ch. In this experiment, a high-speed camera and a Schlieren system are combined to acquire images of the explosive discharge process. The high-speed camera is a Phantom V7.3 Schlieren apparatus and is used to observe the changes in the flow field during the discharge process. The Schlieren apparatus consists of two main reflectors, two small reflectors and a lighting system.

4.1.3 Data Distribution System

The distribution system in this work consists of a vacuum pump, an automatic distribution instrument, a methane supply system and air compressor. The vacuum pump is a model 2X-8GA, the extraction rate is 8 L/s, and the ultimate pressure is 6×10^{-2} Pa. The automatic gas distributor is a rcs2000-b automatic gas distributor system manufactured by Beijing Jinxing Electronics Co., Ltd. This system can realize premixing of any proportion of gas.

4.1.4 Experimental Method

If the ignition source is encountered during the explosion venting process, jet fire is induced. This experiment is carried out in a 55L cylindrical container with a methane concentration of 73.3% UEL. By changing the number of layers of mesh, mesh size and distance between the wire mesh and the vent, different working conditions can be achieved. Experiment group A/B/C mainly studied the flame shape and temperature at different wire mesh layers and meshes, and different distances between the wire mesh and the vent. The experiment group D/E/F mainly studied the flame shape and temperature at different wire mesh layers and meshes, and the different distances from the wire mesh to the first gas cloud. The specific experimental scheme is shown in Table 4.2.

4.2 Results and Discussions

4.2.1 Influence of the Wire Mesh on the Explosion Pressure Behind the Wire Mesh

Influence of the Number of Wire-Mesh Layers on the Release Explosion Venting Pressure

Figure 4.2 shows the changes in the pressure at the measured position and the changes in the maximum explosion pressure for different numbers of wire-mesh layers. Analysis of Fig. 4.2 shows that the wire-mesh structure obviously influences the suppression of the pressure behind the wire mesh. The release pressure and the rate of increase to the maximum explosion pressure decrease with increasing number of

Table 4.2 Research on the wire mesh method for suppressing explosion venting and induced jet fire

Number of wire-mesh layers	Mesh size	Experiment group number	Distance between wire mesh and explosion vent (m)	Experiment group number	Distance between wire mesh and the first gas cloud (m)
0	0	A-1	0	D-1	0
1	60	A-2	0	D-2	0
3	60	A-3	0	D-3	0
5	60	A-4	0	D-4	0
7	60	A-5	0	D-5	0
3	40	B-1	0	E-1	0
3	60	B-2	0	E-2	0
3	80	B-3	0	E-3	0
3	100	B-4	0	E-4	0
3	60	C-1	0.1	F-1	0.1
3	60	C-2	0.2	F-2	0.2
3	60	C-3	0.3	F-3	0.3
3	60	C-4	0.4	F-4	0.4

wire-mesh layers. Without the wire mesh, the maximum explosion pressure at the measured position is 0.65 MPa, and the rate of increase to the maximum explosion pressure is 95.86 MPa·s^{-1}. With a layer of wire mesh, the maximum explosion pressure at the measurement position decreases to 0.56 MPa, which is 13.85% lower than when there is no wire mesh, and the rate of increase to the maximum explosion pressure decreases to 75.03 MPa·s^{-1}, indicating that the addition of wire mesh has a certain inhibitory effect on the explosion venting pressure. Increasing the number of wire-mesh layers causes the surface area of the wire mesh to increase, so that the high temperature gas ejected from the container and the wire mesh come into greater contact. The energy loss from thermal convection between the flame and the wire mesh increases, and the maximum explosion pressure decreases. With 5 wire-mesh layers, the maximum explosion pressure is 0.35, 0.13 MPa lower than the maximum explosion pressure with 3 layers, which is 46.15% lower than when there is no wire mesh, while the explosion pressure increases at a rate of 22 MPa·s^{-1}. With 7 layers of wire mesh, the maximum explosion pressure significantly decreases because with increasing number of wire-mesh layers, when the thickness of the wire mesh reaches a certain level, the explosive energy decreases rapidly due to the effects of heat dissipation and thermal radiation. The degree of total flame energy decreases rapidly with increasing number of wire-mesh layers.

In order to improve the accuracy of the experimental data, repeated experiments were carried out. In the experiment, each group of experiments is repeated three times. In order to describe the experimental results more accurately, the experimental data

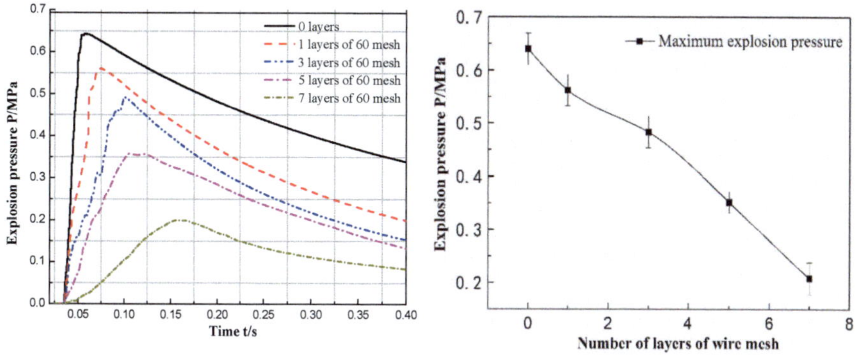

Fig. 4.2 Variations in explosion venting pressure and maximum explosion pressure for different numbers of wire-mesh layers

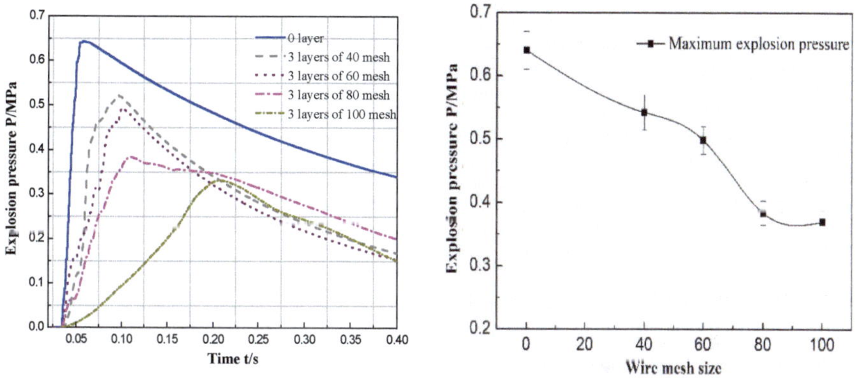

Fig. 4.3 Different mesh sizes with explosion pressure and maximum explosion pressure

and errors of each group are displayed by adding error bars in the Figs. 4.2, 4.3, 4.4, 4.6, 4.8, 4.10, 4.12, and 4.14.

Effect of Mesh Size on the Explosion Pressure Behind the Wire Mesh

Figure 4.3 shows the variation in pressure at a distance 0.4 m from the explosion vent for different mesh sizes. The figure shows that the maximum pressure at the measurement position decreases with increasing mesh size. With 3 layers of size 40 wire mesh, the maximum explosion pressure at the measured position is 0.52 MPa, and the explosion pressure curve obviously fluctuates compared with the curve without wire mesh. When the mesh size increases to 60, the trend of the pressure curve is basically the same as that of the size 40 mesh, the maximum explosion pressure at the measurement position decreases to 0.50 MPa, which is 3.85% lower than that of

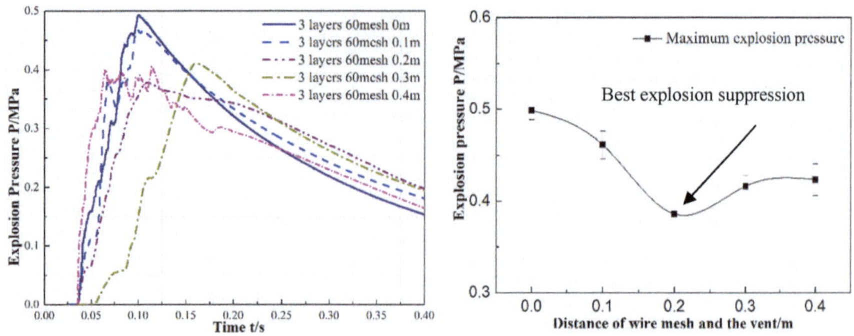

Fig. 4.4 Explosion venting pressure with different wire-mesh positions

the 40 mesh, and the rate of increase to the maximum explosion pressure decreases to 55.89 MPa·s^{-1} respectively. When the mesh size increases to 80, the explosion pressure curve of the measurement position obviously oscillates, and the span of the curve increases. The maximum explosion pressure decreases to 0.38 MPa, which is 26.92% lower than that of the 40 mesh, and the rate of increase to the maximum explosion pressure decreases to 15.89 MPa·s^{-1}. For a mesh size of 100, the maximum explosion pressure decreases to 0.37 MPa, which is 28.85% lower than that of the 40 mesh. It indicates that the mesh size significantly inhibits the pressure of the ejection fire. According to the chain reaction theory of combustion and explosion, the phenomenon of explosion is the result of new radicals being produced and quickly involved in the molecular reaction. Thus, the basic condition for an explosion to occur is that the number of new free radicals produced is greater than the number of free radicals used. As the mesh size increases, the size of the openings in the wire mesh decreases, and the flame and wire mesh are in contact longer. As there is a large temperature difference between the wire mesh and the explosion flame, in the process of contact with the wire mesh, the number of free radicals lost greatly increases, the probability of collisions between free radicals and reaction molecules decreases, and the corresponding explosion capacity decreases, which directly leads to a decrease in the explosion pressure.

Wire-Mesh Position on the Impact of Explosion Pressure Behind the Wire Mesh

The influence of the position of the wire mesh on the explosion pressure behind the wire mesh is shown in Fig. 4.4. When the number of wire-mesh layers is constant, with increasing distance between the wire mesh and the explosion vent, the explosion suppression effect increases; when a certain distance is reached, the pressure reaches a minimum pressure; continuing to increase the distance, the suppression effect weakens; thus, there is an "optimal location for explosion suppression". In this experiment, when the distance reaches 0.2 m, the wire mesh has the best suppression effect. At the critical distance, the pressure wave in front of the wire-mesh structure is blocked by the accumulation of pressure at the wire mesh. After passing through the screen, the pressure is released. However, as the distance between the screen and the release hole continues to increase, the pressure wave is fully accelerated, the flame has largely disappeared, and the release pressure increases.

4.2.2 *Effect of Wire Mesh on Explosion Venting Jet Fire*

Effect of Number of Layers of Wire Mesh on Explosion Venting Jet Fire

(1) Influence of the number of wire-mesh layers on the flame structure

Figure 4.5 shows the Schlieren images that demonstrate the influence of different layers of wire mesh on the flame structure of explosion venting jet fire. Figure 4.5 shows that the spread of the flame is affected to different degrees by the wire mesh. Compared with the schlieren images without the wire mesh, the "mushroom" cloud of gas emitted at 1 ms disperses under the action of the wire mesh, and the "dispersion" effect increases with increasing number of layers of wire mesh. With 7 wire-mesh layers, the amount of gas released through the wire mesh is very small. In addition, when the flame goes through the wire mesh, the flame changes from the original "cylindrical" shape to an irregular shape, and there are different degrees of "bifurcation". The bifurcation effect of the 5-layer wire mesh is the strongest. Increasing number of wire-mesh layers increases the thickness of the wire-mesh structure, and the multilayer wire-mesh structure absorbs part of the heat of the explosion flame, weakening the propagation of the explosion flame. The Schlieren image at 3 ms shows that the 7-layer wire mesh inhibits the blowout flame the most. At 3 ms, the explosion venting jet fires of various working conditions all go through the wire meshes, but the flame of the 7-layer wire mesh is still small compared with those of other groups and fills the whole lens at 5 ms, indicating that the flame propagation speed decreases with increasing number of wire-mesh layers. At 5 ms, for all the of working conditions, the flames are all through the wire mesh, and the turbulent flame area of the 7-layer wire-mesh flame significantly decreases in a short time from a turbulent flame directly into a laminar flame. This result shows that the number of wire-mesh layers impacts the state of the flame propagation; the larger the number of wire-mesh layers is, the more obvious the suppression of the turbulent flame is.

(2) Effect of the number of wire-mesh layers on the ambient temperature outside the vent

Figure 4.6 shows the variations in temperature at a position 0.4 m from the explosion vent for different numbers of wire-mesh layers. Analysis of Fig. 4.6 shows that the number of wire-mesh layers greatly influences the temperature of the jet fire induced by explosion release. Without wire mesh, the maximum temperature of the measured position reaches 903 °C. When a layer of size 60 mesh wire mesh is added,

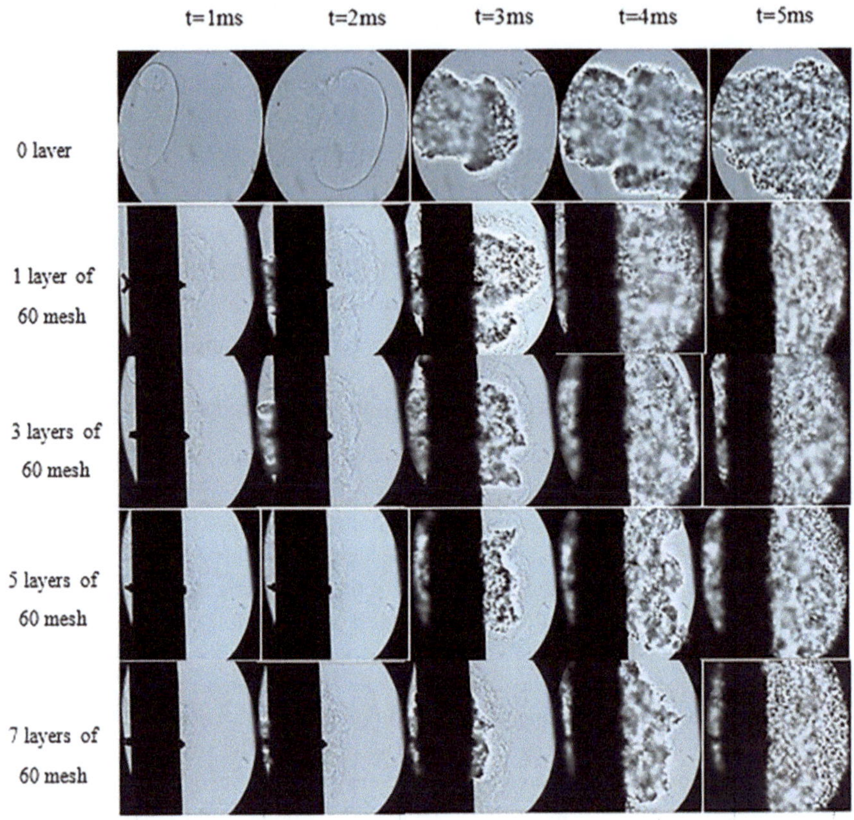

Fig. 4.5 Schlieren image of the influence of different numbers of wire-mesh layers on the flame structure of blowout jet fire

the maximum temperature of the measuring point drops to 816 °C for a decrease of 87 °C, which is 9.63% lower than when there is no wire mesh. When the number of wire-mesh layers reaches 3, due to the increase in the thickness of the wire mesh, the number of gaps between the wire meshes is greater, and the width becomes narrower. As the flame passes through the mesh, the flame is subdivided into strands and enters the mesh holes. Due to the increase in the thickness of the wire mesh, the flame in the gap between the screens remains longer and fully contacts the screen. Because the difference between the temperatures of the wire mesh and the flame is large and the thermal conductivity of the wire mesh is good, the measured position of the highest flame temperature shows a "cliff type" drop, reducing to 503 °C for a decrease of 400 °C.

Increasing the number of wire-mesh layers to 5, the thickness of the wire mesh continues to increase, the flame and the wire mesh are in contact longer, and the flame temperature continues to decrease with a maximum temperature of 306 °C for a decrease of 597 °C, and the maximum temperature is 33.89% of that without

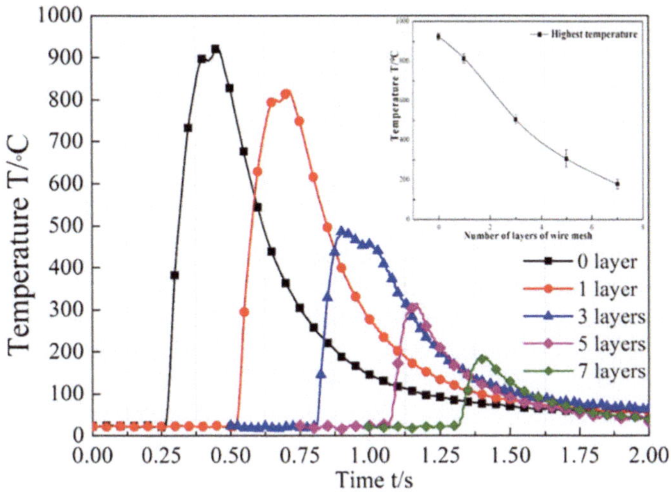

Fig. 4.6 Variations in flame temperature and maximum temperature for different numbers of layers of wire mesh

wire mesh. When there are 7 layers of wire mesh, the thickness of the wire mesh increases further. When the flame passes through the mesh, the flame shuttles in the gap between the wire mesh and reflects and collides back and forth with the wire mesh. The temperature of the flame decreases due to heat loss. The maximum temperature of the flame is further decreased to 178 °C, down to 19.71% when there was no screen. The above experimental engineering results show that the number of wire-mesh layers significantly affects the ambient temperature. In the practical sense, it is possible to prevent secondary disasters by adding a number of layers of wire mesh and making the corresponding explosion relief device.

Effect of Mesh Size on Explosion Venting Jet Fire

(1) Influence of mesh size on flame structure

Studies have shown that the mesh size of a wire mesh significantly impacts the gas explosion intensity. The experimental results show that the 3-layer wire mesh obviously suppresses the jet fire. Therefore, the number of wire-mesh layers is fixed at 3, and the effect of mesh size on the suppression of explosion venting jet fire is studied by changing the mesh size.

Figure 4.7 shows that compared with the Schlieren image without wire mesh, the wire meshes with the same number of layers but different mesh sizes influence the flame structure. At 1 ms, the size 40 mesh screen releases the most gas, and the 100 mesh releases the least. As the mesh size increases, the number of flame bifurcations increases, and the shape of the flame becomes increasingly irregular. The Schlieren diagram shows that the fold of the flame gradually decreases with increasing mesh size, which indicates that the propagation velocity of the flame decreases with increasing mesh size. In conclusion, the different mesh sizes significantly affect the flame propagation and have inhibitory effects on the flame microstructure, that is, the larger the mesh size is, the better the inhibitory effect on the flame propagation is.

(2) Influence of mesh size on the ambient temperature outside the vent

Figure 4.8 shows the changes in the measured position temperature and maximum temperature for different mesh sizes. As seen from Fig. 4.8, different mesh sizes significantly affect the external temperature of the vent. With no wire mesh, the highest temperature of the measured position is 903 °C. For three layers of size 40 mesh, the highest temperature is 663 °C, a decrease of 240 °C, which is 26.58% lower than when there is no screen. When the mesh size is 60, the highest temperature of the flame at the measured position is 468 °C, which is 51.83% of that without the wire mesh. When the mesh sizes are 80 and 100, the maximum temperatures of the measured position decrease to 350 °C and 260 °C, respectively. With increasing

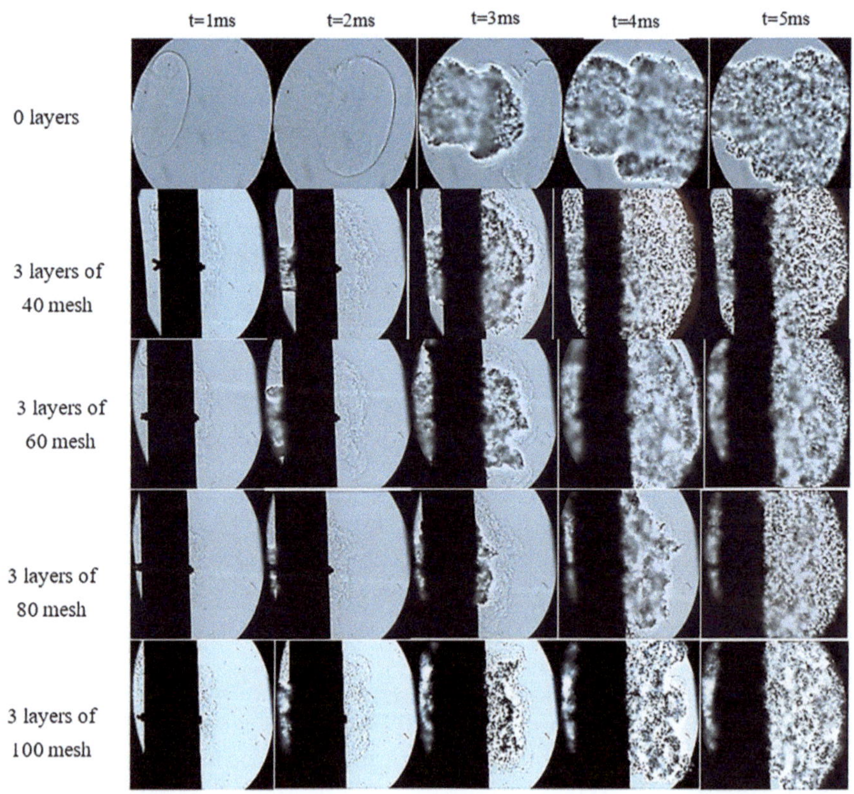

Fig. 4.7 Schlieren diagrams of influence of different mesh sizes on flame structure of explosion venting jet fire

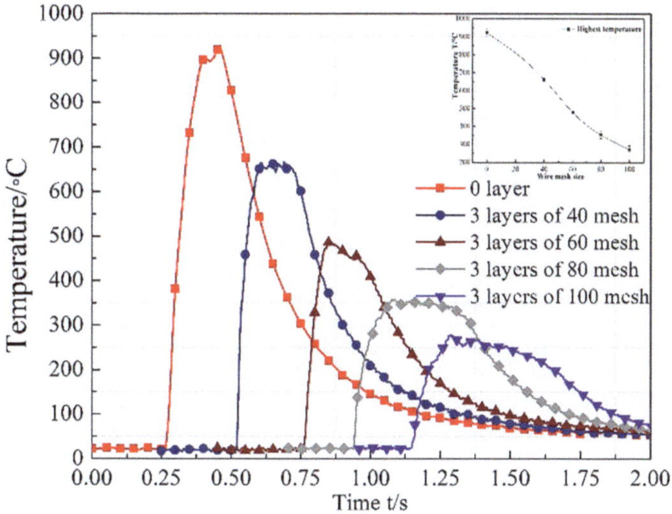

Fig. 4.8 Changes in the temperature outside the vent for different mesh sizes

mesh size, the temperature at the measured position decreases, and the duration of the whole curve lengthens. Although increasing mesh size leads to decreasing ambient temperature outside the vent, the rate of temperature decrease gradually decreases.

Effect of Mesh Position on Explosion Venting Jet Fire

(1) Influence of mesh position on the flame structure

There is not much research on the effect of the wire-mesh position on the suppression of explosion venting jet fire. Based on the above experimental results, this paper chooses 3 layers of 60-mesh as the experimental subject and studies the effect of different mesh positions on the ambient temperature outside the vent by changing the position of the wire mesh (distance from the vent).

Figure 4.9 shows the Schlieren diagram that demonstrate the influence of mesh position on the flame structure. It is found from the figure that the position of the wire mesh has different degrees of influence on the explosion venting jet fire. When the wire mesh is close to the explosion vent, due to the huge disturbance effect of the wire mesh, the flame forms an obvious zigzag-shaped flame front after passing through the wire mesh. This phenomenon is especially obvious at 3 ms. When the distance between the wire mesh and the explosion vent becomes 0.1 m, after the flame passes through the wire mesh, the curvature of the flame front increases, and the surface area of the flame increases. However, a serrated flame is not formed, and a curved flame is formed. The folds become larger and the flame turbulence decreases. When the distance between the wire mesh and the explosion vent is 0.2 m, the flame passing through the wire mesh is already very small, and only part of the flame can penetrate.

The flame wrinkles at the left end of the wire mesh are small and the turbulence is large. But after passing through the wire mesh, the curvature of the entire flame becomes larger, and the flame turbulence behind the wire mesh is greatly reduced, indicating that the wire mesh has a significant inhibitory effect on the explosion venting jet fire. When the distance between the wire mesh and the explosion vent is 0.3 m, the flame passing through the wire mesh increases again, the curvature of the flame front becomes smaller, and the surface area becomes smaller, the inhibitory effect of the wire mesh begins to weaken. When the distance is 0.4 m, the flame area passing through the wire mesh is further increased, and the turbulence of the flame passing through the wire mesh is still weakened. This is because when the distance between the metal wire mesh and the explosion vent is 0.2 m, the precursor pressure wave of the explosion vent jet fire fluctuates violently under the influence of various pressure gradient fields, and it reflects after encountering the wire mesh. Under the action of the reflected wave, the turbulence of the flame propagating forward increases, but the penetrating power is weakened. As the distance increases further, the turbulence of the precursor pressure wave reaching the wire mesh has been reduced. When passing through the screen, the penetration is basically achieved, and the reflected wave is very small. Therefore, the venting jet fire can still have a strong penetrating power under the action of the precursor pressure wave.

(2) Influence of mesh position on the ambient temperature outside the vent

Figure 4.10 shows the variation of the ambient temperature of the wire mesh for different distances from the vent. The figure clearly shows that when the distance between the wire mesh and the vent changes, the temperature outside the vent is affected to a certain extent. When the wire mesh is close to the vent, the highest temperature measured is 468 °C. When the distance is 0.1 m, the maximum flame temperature is 608.43 °C, and when the distance continues to increase to 0.2 m, the temperature of the measured position increases to 653 °C. Then, the temperature gradually decreases; when the distance between the wire mesh and the vent is 0.4 m, the temperature of the measured position decreases to 451 °C. In general, with increasing distance between the screen and the vent, the maximum temperature first increases and then decreases. At 0.2 m, the screen has the worst inhibition effect on the external temperature because at 0.2 m, the area of the flame through the wire mesh decreases. The time of contact between the flame and wire mesh is shorter, the number of free radicals in the wire mesh is lower and the amount heat dissipated is smaller. At greater distances, the amount of flame through the screen increases, but the flame and screen are in contact for a longer time. Therefore, the screen dissipates more heat.

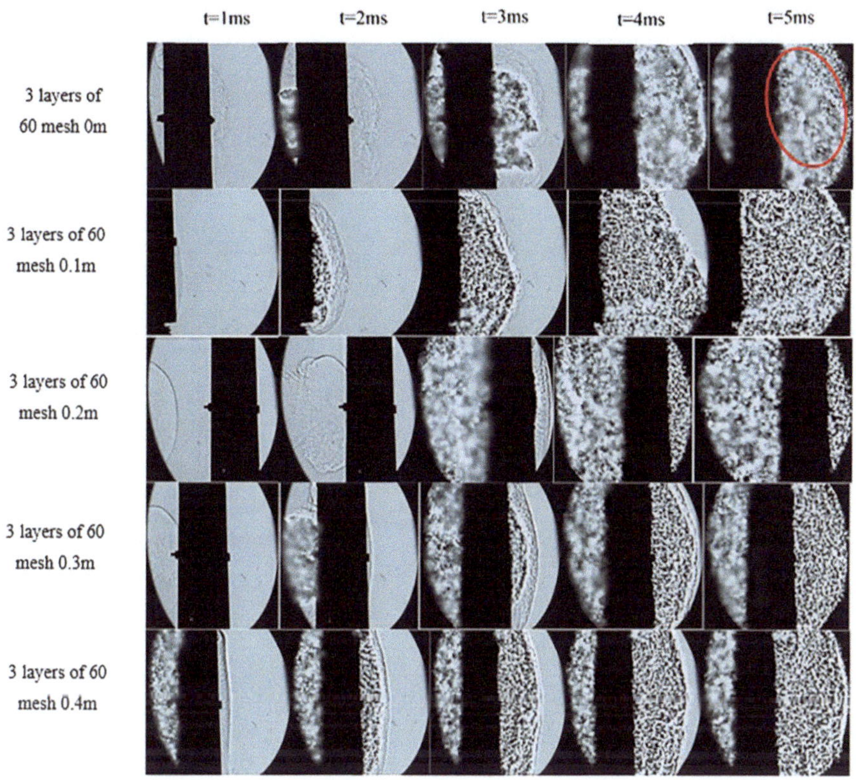

Fig. 4.9 Schlieren diagrams of different mesh position on flame structure of explosion venting

4.2.3 Effect of Wire Mesh on the Chain Reaction of Vapor Cloud Explosion Induced by Venting Jet Fire

Effect of Number of Layers of Wire Mesh on the Chain Reaction of Vapor Cloud Explosion Induced by Venting Jet Fire

(1) Influence of the number of wire-mesh layers on the flame structure

Figure 4.11 shows the Schlieren images that demonstrate the influence of different layers of wire mesh on the flame structure after the explosion jet flame detonates the gas cloud. It is found from Fig. 4.11 that the wire mesh has a significant effect on the flame propagation after the gas cloud explosion induced by the venting jet fire. Compared with no wire mesh, when 1 layer of wire mesh is added, the flame passing through the wire mesh becomes irregular, the surface area of the flame front surface decreases, and the flame front stretches to form a wave or jagged shape. When there are three layers of wire mesh, due to the increase of the thickness of the wire mesh, the intensity of flame turbulence passing through the wire mesh increases

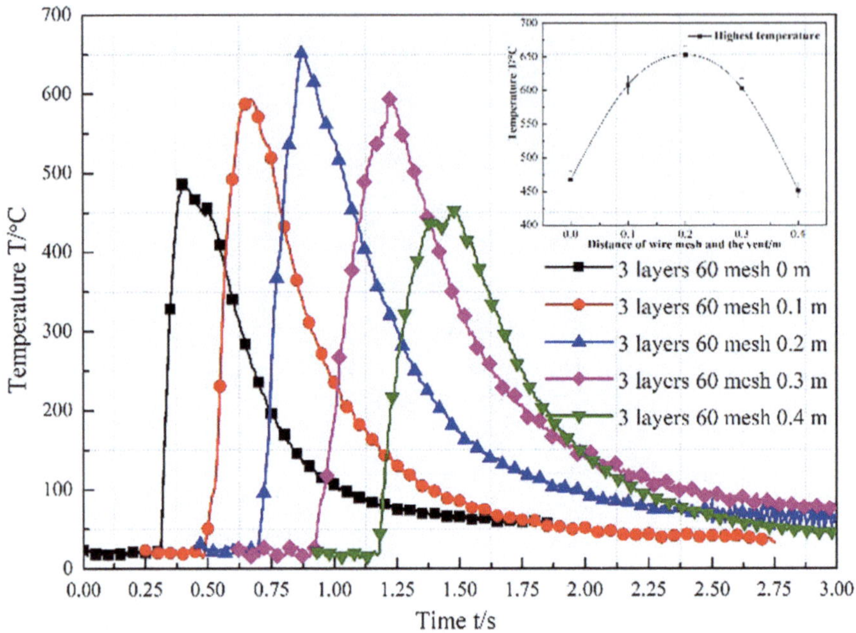

Fig. 4.10 Changes in measured position temperature at different wire-mesh positions

under the action of the pressure field and the separation of the wire mesh. However, the flame surface area is reduced, and there is no jagged flame front, and the front surface becomes smoother instead [2]. When there are five layers of wire mesh, the flame that can pass through the screen becomes very small. Since the venting jet fire detonates the first gas cloud, the overall pressure precursor wave propagates along the upper right, so part of the flame passes through the upper part of the wire mesh. This phenomenon is particularly obvious when the number of layers of the wire mesh is 7 layers. In general, with the increase of the number of metal wire mesh layers, the more obvious the effect of suppressing the flame propagation after the explosion of the first gas cloud detonated by the venting jet fire.

(2) Effect of the number of wire-mesh layers on the ambient temperature the ambient temperature after gas cloud explosion

Analyzing Fig. 4.12, it is found that with the increase of the number of metal wire mesh layers, the better the suppression effect of the metal wire mesh on the vent flame. Compared with no wire mesh and 1-layer wire mesh, when there are 3 layers of wire mesh, the temperature after the flame of the first gas cloud explosion passes through the wire mesh and drops to 490 °C, which is about 400 °C lower than without the wire mesh. It can be seen that the metal wire mesh with enough layers can effectively inhibit the flame propagation after the explosion of the gas cloud caused by the vent flame, and it is also the point where the temperature drop is the largest. When the

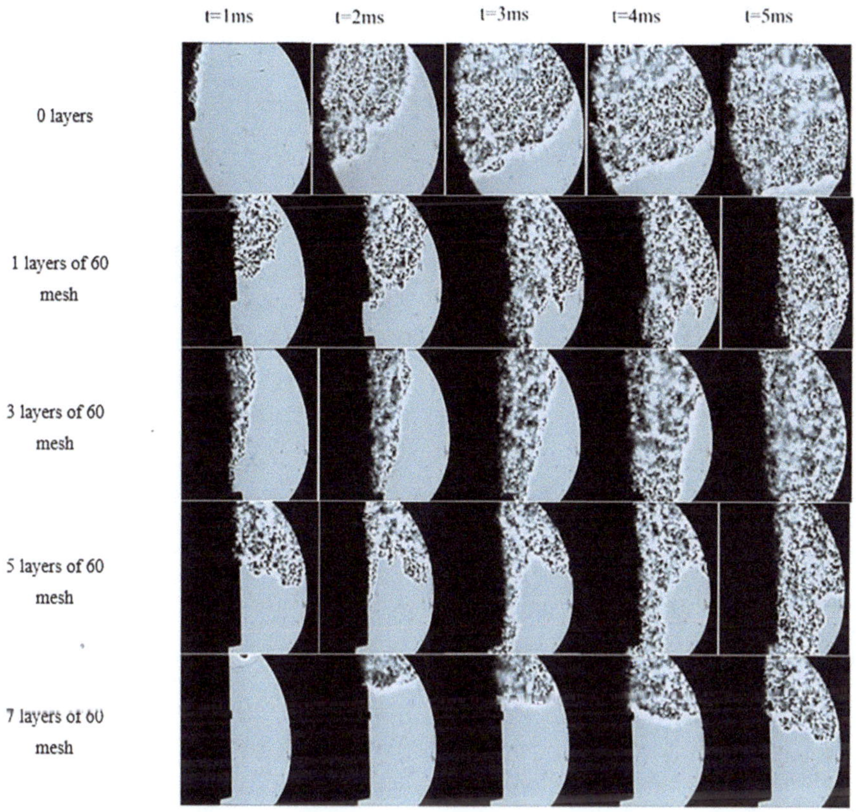

Fig. 4.11 Schlieren image of the influence of different numbers of wire-mesh layers on the chain reaction of vapor cloud explosion induced by venting jet fire

number of screen layers increased to 5 and 7, the flame temperature dropped to a very low level, but the drop began to decrease.

Effect of Mesh Size on the Chain Reaction of Vapor Cloud Explosion Induced by Venting Jet Fire

(1) Influence of mesh size on the flame structure

Studies show that the mesh size of the wire mesh will have a significant impact on the intensity of gas cloud propagation and explosion. The three-layer wire mesh has obvious effect on suppressing the explosion intensity after the gas cloud explosion induced by the venting jet fire. Therefore, the number of fixed screen layers is 3 layers. And the effect of mesh size on suppressing the gas cloud explosion induced by the venting jet fire is studied.

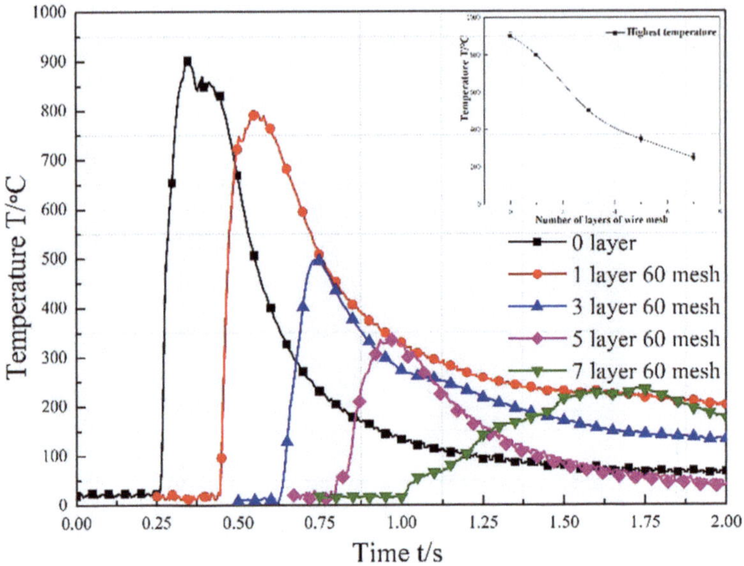

Fig. 4.12 Variations in flame temperature for different numbers of layers of wire mesh

Figure 4.13 shows the Schlieren images that demonstrate the influence of mesh size on the flame structure after the explosion jet flame detonates the gas cloud. Compared with no wire mesh, after adding 3 layers of 40 mesh wire mesh, the multi-layer wire mesh structure absorbs part of the heat of the explosion flame and weakens the propagation of the explosion flame. The surface area of the flame becomes smaller and the flame front becomes irregular. As the mesh number of the wire mesh increases, the area of flame passing through the wire mesh becomes smaller in the same time, and shows a phenomenon of "more flames in the upper half and less in the lower half". When the mesh number of the wire mesh is 60, as the aperture of the wire mesh becomes smaller, the barrier rate of the wire mesh becomes larger. Under the interaction of the precursor pressure wave and the wire mesh, turbulence is caused, which makes the flame turbulence greater, but the flame surface area remains smaller. This phenomenon still exists in 80 mesh and 100 mesh. Therefore, for storage tank sites where explosion venting may occur, the wire mesh explosion-proof method can restrain the occurrence of accidents in this life to a certain extent.

(2) Effect of mesh size on the ambient temperature the ambient temperature after gas cloud explosion

Figure 4.14 shows the Schlieren images that demonstrate the influence of mesh size on the flame structure after the explosion jet flame detonates the gas cloud. The wire mesh structure of different meshes can obviously inhibit the flame temperature propagation after the first gas cloud explosion induced by the venting jet fire. In addition, the inhibition effects of different meshes of wire mesh are different. 100 meshes have the best inhibition effect. The maximum temperature directly drops from

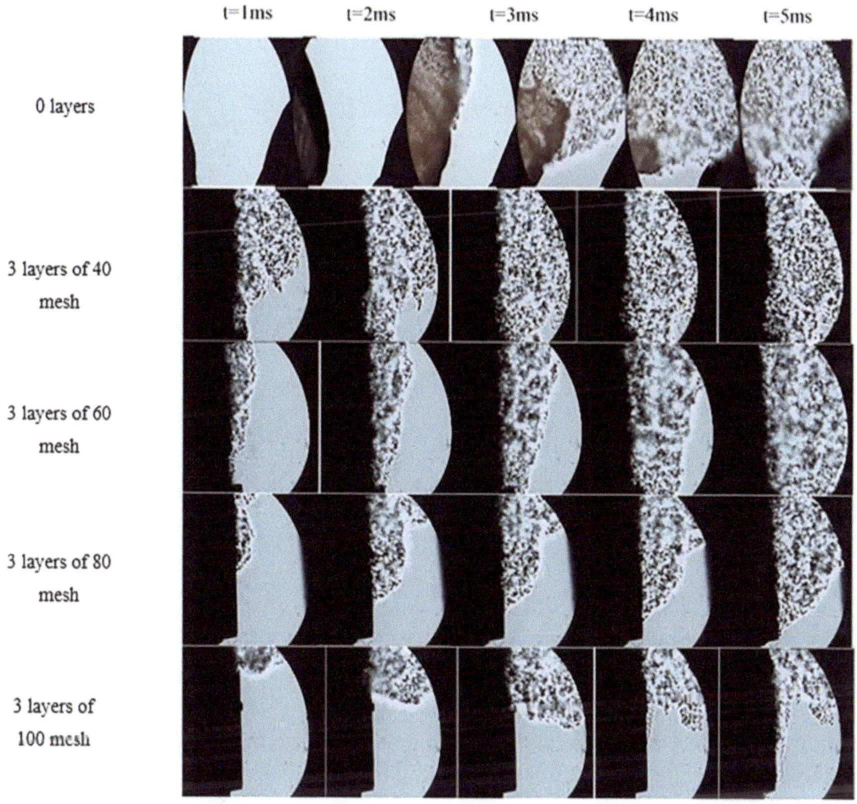

Fig. 4.13 Schlieren image of the influence of mesh size on the chain reaction of vapor cloud explosion induced by venting jet fire

903 to 128 °C, and the drop reaches 780 °C. Compared with wire mesh with different layers, wire mesh with different meshes have better flame suppression effect. This is because as the mesh number of the wire mesh increases, the gap of the wire mesh will increase accordingly. When the flame passes through the wire mesh with more meshes, it will take longer to shuttle inside the wire mesh, which will occur between the wire mesh and the wire mesh. The reflection and attenuation phenomenon is more obvious. Since the metal wire mesh has a better heat conduction effect and a large temperature difference with the flame, the heat can be quickly discharged after full contact with the flame, which directly causes the temperature of the flame to drop.

Effect of Mesh Position on the Chain Reaction of Vapor Cloud Explosion Induced by Venting Jet Fire

(1) Influence of mesh position on the flame structure

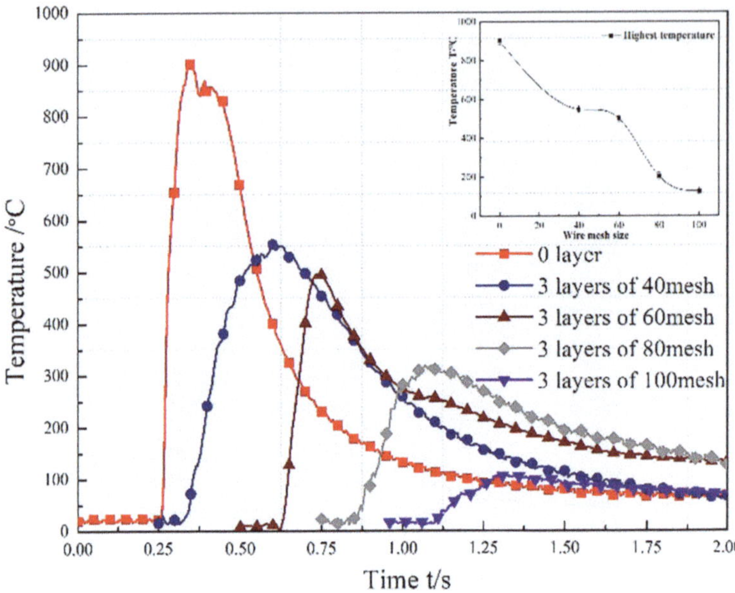

Fig. 4.14 Variations in flame temperature and maximum temperature for different size of wire mesh

Figure 4.15 shows the Schlieren images that demonstrate the influence of mesh size on the flame structure after the explosion jet flame detonates the gas cloud. When the position of the wire mesh changes, the suppression effect on the flame propagation is not obvious, but the suppression effect of different positions is different, but the difference is not big. When the wire mesh is close to the first gas cloud, there is less flame passing through the wire mesh, the degree of wrinkle of the flame front is smaller, and the flame front does not appear jagged. When the distance between the wire mesh and the first gas cloud increases to 0.1 m, the surface area of the flame front after passing through the wire mesh increases, the flame front wrinkles more, the turbulence degree is large, and a jagged flame appears. When the distance between the wire mesh and the first gas cloud increases to 0.2 m, the shape of the flame after passing through the wire mesh does not change significantly, the flame front is smooth, the flame wrinkles are further reduced, and the degree of flame turbulence increases. As the distance continues to increase, although the shape of the flame changes, it does not change much. But when the distance is 0.4 m, the flame passing through the wire mesh becomes very few, the folds of the flame passing through the wire mesh become larger, the degree of turbulence becomes smaller, and the flame has a more obvious inhibitory effect.

(2) Effect of mesh size on the ambient temperature the ambient temperature after gas cloud explosion

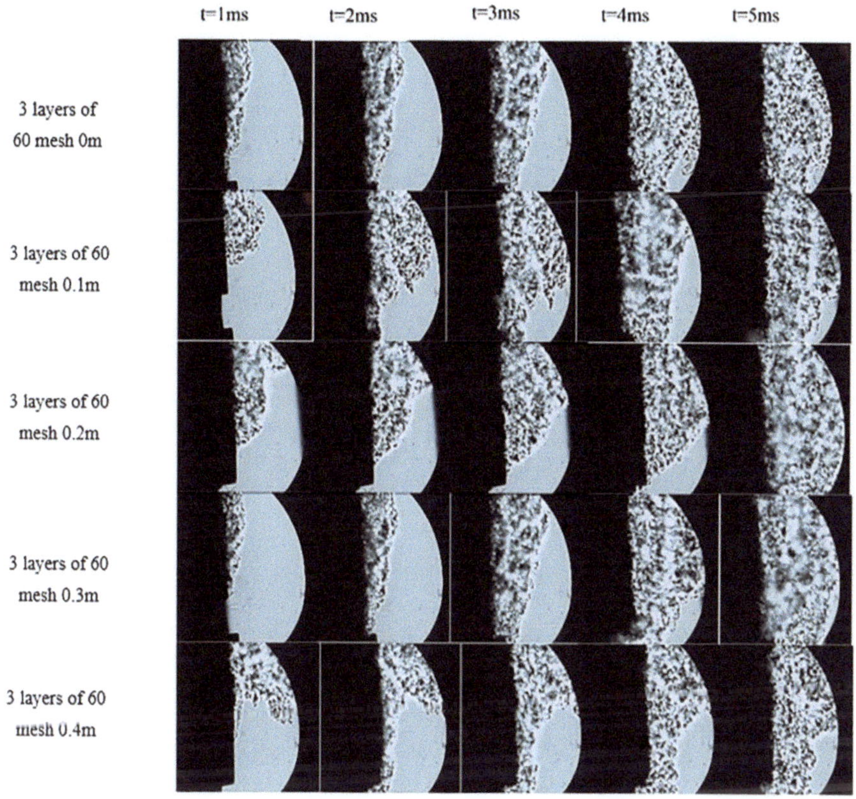

Fig. 4.15 Schlieren image of the influence of mesh position on the chain reaction of vapor cloud explosion induced by venting jet fire

It can be seen from Fig. 4.16 that as the distance between the wire mesh and the first gas cloud changes, the temperature of the flame gradually increases. This shows that as the distance increases, the effect of the metal wire mesh on suppressing the flame temperature of the first gas cloud caused by the venting jet fire to explode gradually becomes worse. When it is close to the first gas cloud, the highest flame temperature is about 500 °C, and when the distance increases to 0.1 m, the highest temperature increases to 600 °C. After that, as the distance increases, the temperature increases but the difference is not obvious. According to the mall effect theory, when the flame front is in contact with the wire mesh, the chance of the flame colliding with the wire mesh structure increases. The number of free radicals destroyed by the mall effect increases. Under the segmentation of the metal grid, many flames will be divided into small fire groups. However, this kind of fireball will lose its thermal balance under the turbulence generated during the contact between the flame precursor pressure wave and the wire mesh. The direct result is that the flame extinguishes and the flame temperature decreases. When the wire mesh is close to the gas cloud, the flame propagation rate is faster, and the turbulence generated by the contact between the

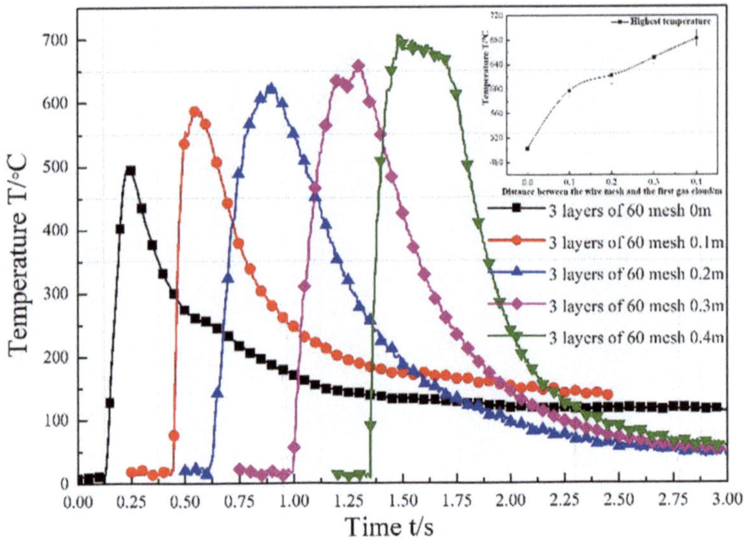

Fig. 4.16 Variations in flame temperature and maximum temperature for different wire mesh position

flame and the wire mesh is greater, and more flames are extinguished and the heat loss is greater. As the distance between the metal wire mesh and the first gas cloud increases, the flame propagation rate decreases, and when it comes into contact with the metal wire mesh, the turbulence generated becomes smaller and the heat loss is also smaller.

The minimum ignition energy is obtained by the test method currently, and the calculation formula of the minimum ignition energy is fitted through experimental data. At the same time, according to ASTM E582-07, the measurement of ignition energy needs to be repeated 5 times. The minimum ignition energy of methane is relatively small. In this experiment, the ignition energy is 20 J, and the voltage is 220 V to ignite the methane premixed gas inside the container. The vented flame can ignite the gas cloud and explode after passing through the wire mesh. It shows that the vent flame has reached the minimum ignition energy of methane premixed gas. There are many factors affecting the minimum ignition energy of methane, including methane concentration, ambient temperature, etc., which have a significant impact on the minimum ignition energy.

4.3 Conclusions

In this paper, mechanisms of suppressing the explosion and the jet fire when the venting jet fire induces single gas cloud explosion and gas cloud chain explosion by metal wire mesh are studied. The influences of the number of mesh layers, mesh size and screen position on the explosion pressure, flame structure of the ejection fire and flame temperature are explored. The relevant conclusions are drawn as follows.

(1) As the number of mesh layers and the mesh size increase, the venting pressure decreases.
(2) The wire mesh has a restraining effect on the flame structure and the propagation rate. As the number of mesh layers and mesh size increase, the strength of the flame passing through the screen weakens, and the flame propagation speed slows.
(3) As the number of screen layers and mesh size increase, the temperature of the external environment of the wire mesh decreases.
(4) For a single gas cloud, the suppression effect is most obvious when the distance between the wire mesh and the explosion vent is 0.2 m. In the case of double gas clouds, the suppression effect is the most obvious when the distance between the screen and the first gas cloud is 0.4 m.

References

1. Wang ZR, Lu YW, Cao XY, Yu Y, Jiang JC, Jiao F, Ma SC. Wire-mesh inhibition of jet fire induced by explosion venting. J Loss Prevent Process Ind. 2021;70:104408.
2. Cheng GB, Zhang XD, Li JX, et al. Experimental study on the physical mechanism of premixed flame propagation in obstacle pipeline. J Saf Environ. 2014;14(1):56–60.

Chapter 5
Research on the Synergy Inhibition of Ultrafine Water Mist and Metal Wire Mesh on the Syngas Explosion

5.1 Experimental Apparatus and Procedures

Figure 5.1 presents the structural diagram of experiment apparatus for the synergy inhibition of syngas explosion using the ultrafine water mist and metal wire mesh. The main structure was introduced in detail in the previous work [1]. This experimental apparatus was mainly equipped with a metal mesh flame resistance system and a spray system. The flame resistance system consisted of two wire mesh press plates, the multi-layer metal wire mesh (the stainless steel) and four support rods. The metal wire mesh was made of 304 stainless steel material and its thermal conductivity was 16.3 W m^{-1} K^{-1} at 100 °C. The melting point could reach 1457 °C. The wire mesh with determined layer and mesh numbers was placed inside the press plate, and four support rods were adopted to support it. The wire mesh was compressed tightly and the height of press plate was adjusted through bolts (Effective area: $H \times D = 100$ mm $\times$ 100 mm). To ensure the same compression strength, a metal compression ring with a certain height was installed on the internal support rod of press plate and tightened with bolts. The structure size of press plates was designed and processed based on the cross-sectional area of explosion vessel to achieve close integration. Meanwhile, the gap of wall surface and the press plate was blocked by the vacuum sealing mud. The spray system included a water storage tank, a solenoid valve, one-way valve and a fine nozzle, as shown in Fig. 5.1. The fine nozzle was installed at the bottom end of vessel top flange. The water storage tank was filled with the deionized water and its internal pressure was determined by a precision pressure gauge. By controlling the opening and closing of the solenoid valve through the program, the different spray time was achieved.

During the experiments, the metal wire mesh was adopted to prevent the flame propagation of syngas explosion and the synergy effect with spray mist was conducted. A high-speed camera with 2000 fps acquisition frequency was adopted to record the dynamic propagation characteristics of flame under the synergy effect. Two high-frequency pressure transmitters with 200 kHz acquisition frequency and

Z. Wang and X. Cao, *Flame Resistance Characteristics and Its Influence Mechanism*,
https://doi.org/10.1007/978-981-95-6119-3_5

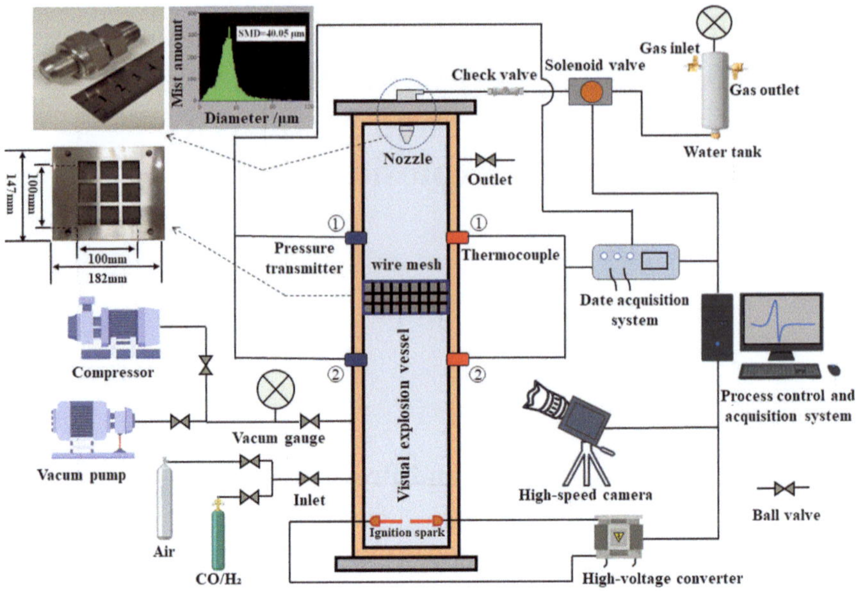

Fig. 5.1 Schematic diagram of explosion resistance/inhibition experiment apparatus

two high-precision thermocouples with 0–2000 °C acquisition range were installed at the upper and lower ends of flame resistance system to collect pressure and flame temperature data. The vessel was vacuumed before the experiment, and syngas and air were supplied through the gas cylinder. The concentration of premixed gas inside the vessel was determined through the Dalton's partial pressure law. A high-voltage discharge method (U = 8 kV) was adopted to ignite premixed gas. The sequence of ignition, spray and data acquisition was realized through the program control and acquisition system (Acquisition frequency: 200 kHz/per channel).

Three syngas concentrations were selected during the experiments (H_2: CO = 1:1; c = 15, 20 and 25%). Based on the previous experimental data, the structural parameter of wire mesh was determined [Layer number (n = 25) and Mesh number (m = 40)]. The flame resistance system was installed at a distance of 650 mm from the bottom end of vessel, which ensured the full development of syngas explosion flame [2]. The diameter of droplet in the mist generated by nozzle with TG-SS 0.3 type was measured using a phase doppler particle analyzer (PDPA). Under 0.6 MPa spray pressure, the droplet diameter [Sauter mean diameter (SMD) = 40.05 μm] was measured at a distance of 30 mm below the nozzle. By changing the spray time (t_{sp} = 0, 200, 300, 400, 500 and 600 ms), the spray amount above the flame resistance system was increased. The premixed gas was ignited after standing for 100 ms. The flame accelerated upward propagation from the bottom end of vessel. At this moment, the program control system began to collect and store data. In order to analyze and validate the accuracy of the results, the experiment for each setting was repeated three times. The maximum deviation values of p_{max}, $(dp/dt)_{max}$, v_{max} and T_{max} were 5.8%, 7.6%, 6.3% and 5.2%, respectively.

5.2 Results and Discussions

5.2.1 *Analysis of Explosion Resistance/inhibition Characteristics*

To clarify the synergy effect, the correspondence between flame propagation process and pressure change under the synergy inhibition were analyzed. Figure 5.2 illustrates the correspondence between the pressure rise and flame propagation change on both sides of flame resistance system under the synergy effect ($c = 25\%$, $t_{sp} = 200$ ms, $P_{sp} = 0.60$ MPa). It clear that the synergy effect failed under this condition. The flame developed in a "spherical" structure after the premixed gas was ignited, but the pressure history at the lower end did not present an obvious increase trend. After 6.00 ms, the flame developed into a "finger" structure and accelerated propagation, accompanied by an obvious increase in the rate of pressure rise. At 13.25 ms, the velocity reached the first peak value ($v_1 = 42.5$ m/s). Due to the obstruction of flame resistance system, the velocity history appeared a downward trend. After approaching the wire mesh, the flame appeared the secondary acceleration process and velocity history reached the second peak value at 17.25 ms ($v_2 = 40.5$ m/s). The rate of pressure rise at the lower end was evidently decreased after passing through the wire mesh, and even appeared a larger negative value. Afterwards, the pressure was intensified and raised again due to the continuous heat accumulation at the lower end. At 28.50 ms, the rate of pressure rise reached its peak value [$(dp/dt)_{max} =$ 44.6 MPa/s]. Subsequently, the flame at the lower end was gradually extinguished and the corresponding pressure rise rate was also continuously decreased. At 33.75 ms, the pressure at the lower end reached its peak value ($p_{max} = 0.35$ MPa). T_{max} of corresponding flame at the lower end was 1741 °C. Nevertheless, the pressure history showed an upward trend although the flame did not propagate to the upper end at 10.00 ms. At 22.25 ms, it passed through the flame resistance system after the flame stayed for 5.00 ms inside the wire mesh and ignited premixed gas of upper end. Compared with the lower end, the propagation velocity of flame at the upper end was evidently accelerated. Under the action of shock wave, the pressure of upper end appeared an obvious increase. At 19.60 ms, the (dp/dt) history exhibited the first peak value [$(dp/dt)_{max(1)} = 18.9$ MPa/s]. And the velocity history reached its peak value at 24.50 ms ($v_{max} = 157.4$ m/s). With the acceleration propagation of flame and the rapid accumulation of heat at the upper end, the rate of pressure rise was increased greatly. Especially the (dp/dt) history appeared the second peak at 26.75 ms [$(dp/dt)_{max(2)} = 167.7$ MPa/s]. Afterwards, the rate of pressure rise was sharply decreased and its change got more significant compared with the lower end, and even reaching −79.8 MPa/s in $(dp/dt)_{min}$. Simultaneously, the p_{max} was larger and appeared earlier than that of lower end ($t = 29.00$ ms, $p_{max} = 0.55$ MPa). Besides, the flame temperature was also evidently increased and T_{max} reached 1953 °C. It fully explains that the explosion reaction at the rear end of flame resistance system could be evidently enhanced as the inhibition failed.

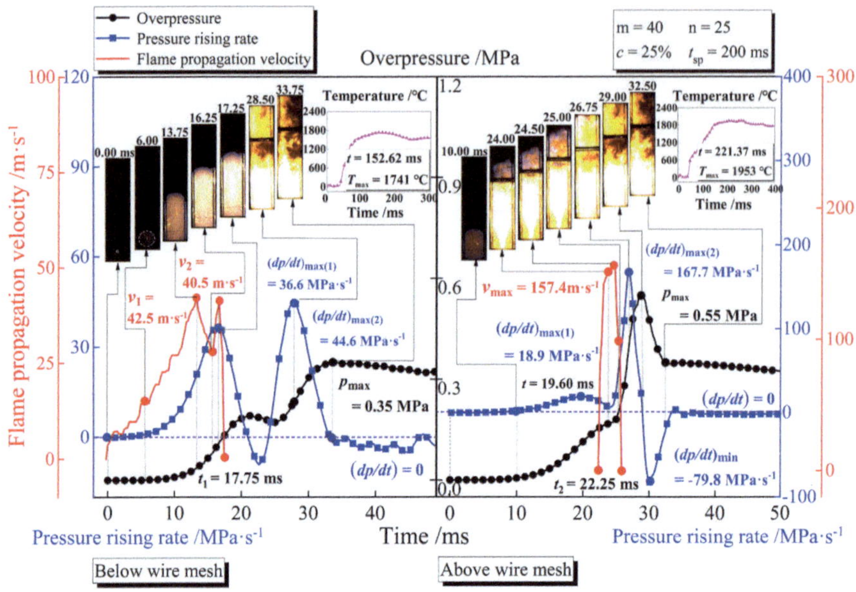

Fig. 5.2 Correspondence between explosion parameters under synergy failure (c = 25%, t_{sp} = 200 ms, P_{sp} = 0.60 MPa)

Figure 5.3 presents the change trend of pressure on both ends of flame resistance system under synergy action as the flame propagated (c = 25%, t_{sp} = 600 ms, P_{sp} = 0.60 MPa). Under this condition, the synergy inhibition was successful. During the development stage of "spherical" flame, the moment of pressure rise did not change compare with the failure. Nevertheless, the flow field near the lower end of flame resistance system showed varying degrees of disturbance due to the increase of spray time, leading to a slight increase in the velocity during the later stage of acceleration [3]. And the moment for reaching the first peak value of velocity history was slightly advanced (t = 12.75 ms). The second peak value also appeared earlier (t = 17.00 ms). Moreover, the corresponding two peak values also presented a slight increasing trend of change. Nevertheless, the *(dp/dt)* history at the lower end only showed an acceleration process as the flame resistance was successful compared with the failure, and *(dp/dt)* history was obviously reduced [$(dp/dt)_{max}$ = 19.6 MPa/s]. The value was evidently decreased although the appearance moment of p_{max} was advanced. Besides, T_{max} was also decreased (T_{max} = 1555 °C). There was still a clear upward process in the pressure history at the upper end although the flame did not pass through the wire mesh. The *(dp/dt)* and p_{max} were greatly reduced compared with the failure. However, the change in *(dp/dt)* history was only 1.9 MP/s and there was a slight change in p_{max} than the lower end as the synergy effect was successful. Besides, the temperature history did not show a significant upward trend (T_{max} = 90 °C). This fully explains that the increase of spray amount could enhance the synergy effect, but the pressure still presented an obvious upward trend.

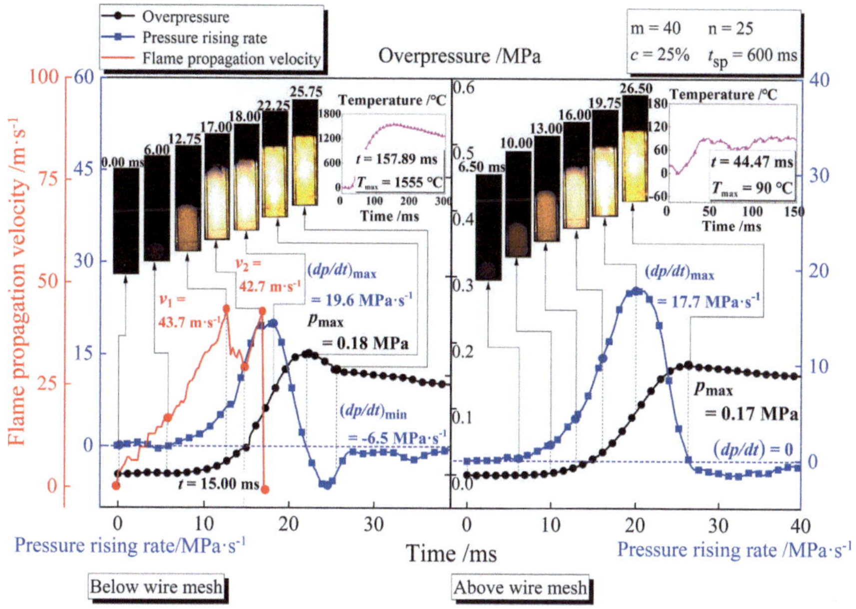

Fig. 5.3 Correspondence between explosion parameters under synergy success ($c = 25\%$, $t_{sp} = 600$ ms, $P_{sp} = 0.60$ MPa)

5.2.2 *Analysis of Flame Resistance Characteristics*

Figure 5.4 shows the influence of spray time on flame propagation process under synergy effect ($c = 25\%$, $P_{sp} = 0.60$ MPa). The propagation velocity of flame at the lower end was slightly increased and presented the light blue with the increase of spray time. The moment for entering the flame resistance system was evidently advanced and its front structure appeared the obvious contraction and acceleration. This was because that the increase in spray time resulted in the enhancement of flow field turbulent disturbance at the upper end, further causing the increase in the flow field disturbance near the lower end of wire mesh due to its penetrability. Meanwhile, the reflection effect of wire mesh on the pressure wave was also enhanced, thereby causing the change in flame structure and a certain increase on the flame propagation velocity. Under the condition no mist, the premixed gas of upper end was ignited firstly and propagated upwards with the light blue flame after stagnation inside the flame resistance system. Then the flame passed through the wire mesh and ignited the premixed gas of upper end under 200 ms spray time at 23.00 ms. With the increase of spray time, the time for passing through the wire mesh was evidently prolonged. For 500 ms spray time, the flame finally passed through the wire mesh. Especially, the flame failed to ignite the premixed gas as $t_{sp} = 600$ ms. It clear indicates that the mist produced by spray method had a good inhibition effect on the syngas explosion, and the effect was enhanced with the increase of spray time.

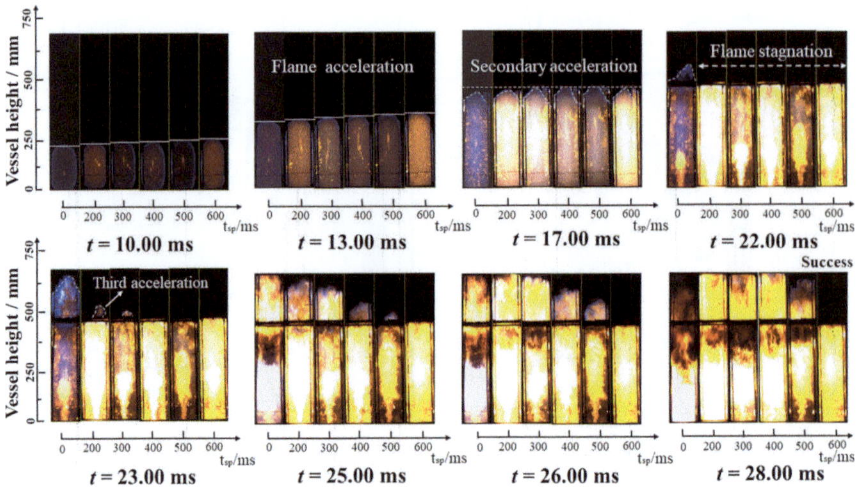

Fig. 5.4 Comparisons of flame propagation processes under different spray times (c = 25%, P_{sp} = 0.60 MPa)

Figure 5.5 shows the effect of spray time on the velocity history under synergy inhibition (c = 25%, P_{sp} = 0.60 MPa). And it exhibited the same change regular as the flame propagated. In the early stage of flame propagation, the velocity history at the lower end did not appear the obvious difference. With the increase of spray time, the velocity history in the acceleration stage was increased but it was not significant. Due to the obstruction of flame resistance system, the velocity history appeared a downward trend after reaching the first peak value. Afterwards, the flame accelerated along the central axis as it approached the wire mesh due to the obstruction of circumferential press plate, thereby exhibiting the second acceleration process. Meanwhile, the velocity entering the wire mesh was evidently increased with the rise of spray time. Subsequently, the flame stagnated inside the flame resistance system and the time was also increased from 2.50 to 7.25 ms. After passing through the wire mesh, the velocity history showed a change trend of increasing first and then decreasing. This was mainly influenced by the combined effect of turbulence disturbance of wire mesh and the heat absorption of mist on the flame [4].

Figure 5.6 illustrates the velocity histories at the both ends corresponding to three kinds of syngas concentrations under synergy effect (t_{sp} = 200 ms, P_{sp} = 0.60 MPa). As the syngas concentration decreased, the slope of velocity history at the both ends of wire mesh were reduced sequentially. Compared with the effect of spray time, the reduction extent was more significant. The velocity history at lower end emerged two acceleration upward processes, but the two peaks were obviously decreased and their adjacent time was prolonged continuously. Simultaneously, the stagnation time was evidently prolonged inside the flame resistance system (from 5.25 to 13.50 ms). Besides, the velocity corresponding to 15% syngas concentration under the action of mist was the smallest after passing through the flame resistance system. Especially

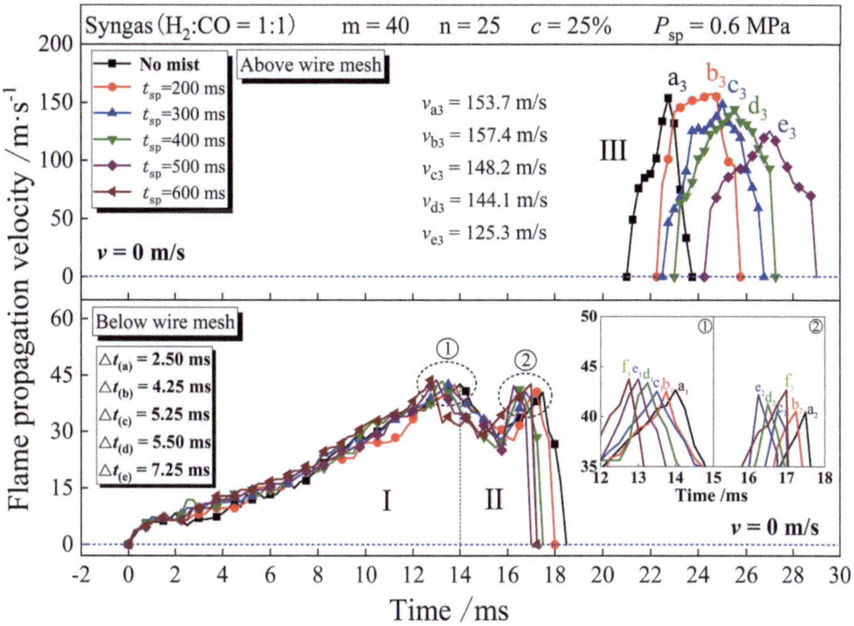

Fig. 5.5 Comparisons of velocity histories under different spray times ($c = 25\%$, $P_{sp} = 0.60$ MPa)

the v_{max} was decreased from 157.4 to 23.1 m/s. And the time for reaching the top end was increased. It presented the same change trend with the decrease of syngas concentration under different spray amounts, as shown in Table 5.1. Besides, the synergy effect was also enhanced with the increase of spray time.

5.2.3 Analysis of Explosion Pressure

Figure 5.7 shows the effect of spray time on the pressure at the both ends of wire mesh under synergy inhibition ($c = 25\%$, $P_{sp} = 0.60$ MPa). Compared with the lower end, the pressure at the upper end was evidently increased under the failure condition. Under 200 ms spray condition, the pressures at the both ends were slightly higher than without spray. However, the pressure was continuously decreased as the spray time increased. And the decrease extent at the upper end was more significant. Especially the synergy effect was successful under 600 ms spray time and the pressures at the two ends were greatly reduced. It was reduced to 0.17 MPa and 0.18 MPa respectively. At this moment, the pressure at lower end was slightly higher than the upper end. This indicates that the mist generated by the spray method could effectively inhibit the propagation of syngas explosion flame, and the increase in spray amount could greatly reduce its explosion intensity. However, the less spray amount had a certain enhancing effect on the syngas explosion.

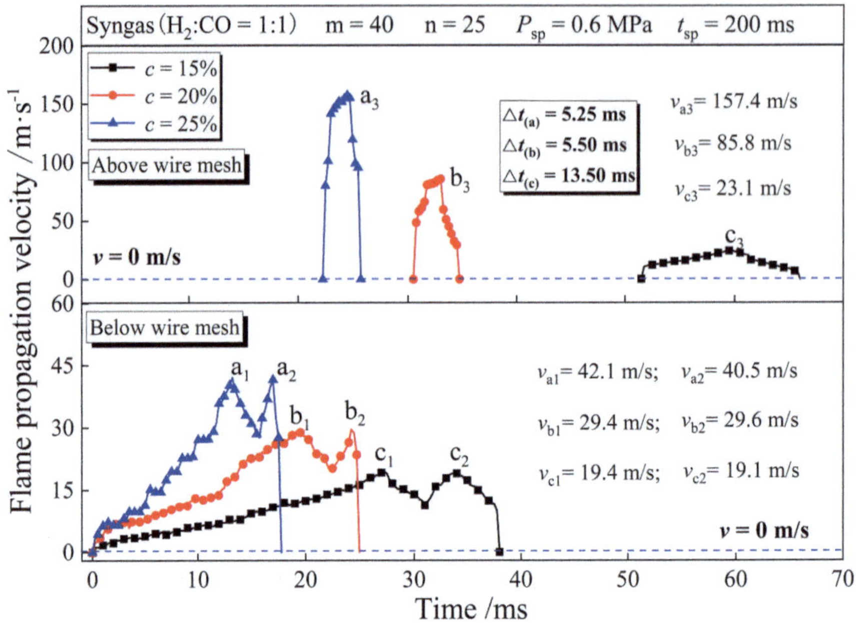

Fig. 5.6 Comparisons of velocity histories under different syngas concentrations (t_{sp} = 200 ms, P_{sp} = 0.60 MPa)

Table 5.1 v_{max} corresponding to different spray times and syngas concentrations

v_{max}/m s^{-1}	Syngas concentration								
	$c = 15\%$			$c = 20\%$			$c = 25\%$		
t_{sp}/ms	Below		Above	Below		Above	Below		Above
0	25.9	23.6	46.9	32.4	33.7	97.1	42.7	40.4	153.7
200	19.4	19.1	23.1	29.4	29.6	85.8	42.8	41.0	157.4
300	16.4	16.2	—	28.1	28.7	77.7	43.1	41.5	148.2
400	/	/	—	27.5	27.1	74.5	43.3	42.2	144.1
500	/	/	—	25.9	25.6	—	43.6	42.6	125.3
600	/	/	—	/	/	—	43.7	43.1	—

"—" Represent "No flame propagation"; "/" Represent "No experiments"

"—" Represent "No flame propagation"; "/" Represent "No experiments"

Figure 5.8 presents the effect of syngas concentration on the pressure histories at the both ends of wire mesh under synergy inhibition (t_{sp} = 200 ms, P_{sp} = 0.60 MPa). The pressure at the both ends was slightly increased at a concentration of 25% syngas under 200 ms spray time compared with the absence of mist. As the syngas concentration decreased, the inhibition effect was enhanced and the decrease extents of

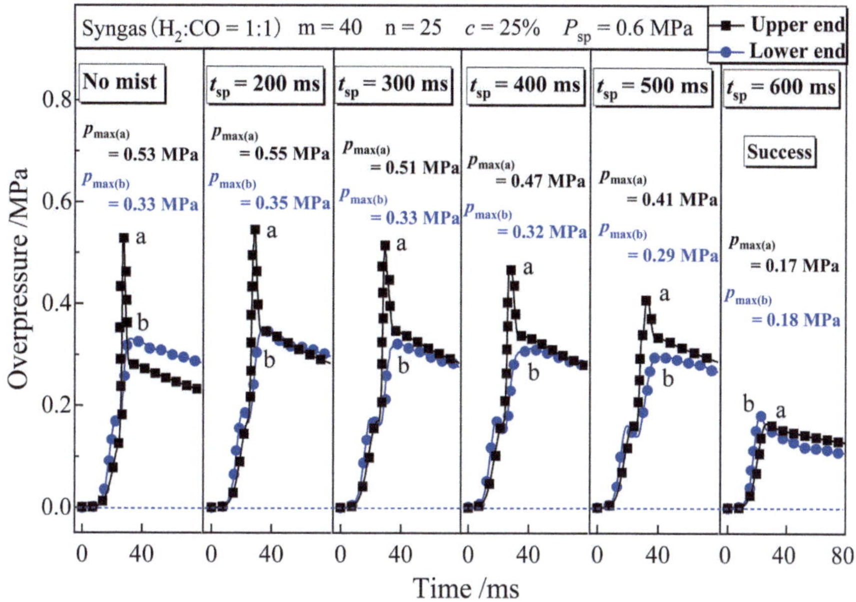

Fig. 5.7 Comparisons of pressure histories under different spray times ($c = 25\%$, $P_{sp} = 0.60$ MPa)

pressures at the both ends were evidently increased. Table 5.2 illustrates the influence of spray time on p_{max} under different syngas concentrations. The lower part of red line in the table represented the working condition corresponding to synergy inhibition success. It clear that the synergy effect was constantly enhanced with the decrease of syngas concentration and the increase of spray time.

5.2.4 *Analysis of Pressure Rising Rate*

Figure 5.9 presents the effect of spray time on *(dp/dt)* histories at the upper and lower ends under synergy inhibition ($c = 25\%$, $P_{sp} = 0.60$ MPa). Its influence regular was corresponding to the change in the pressure history. The *(dp/dt)* history emerged two peaks and the second peak value at the upper end was larger as the synergy inhibition failed. And the second peak value at the two ends presented a slight increasing trend as $t_{sp} = 200$ ms. Then the peak values were decreased with the increase of spray time. Especially the *(dp/dt)* history only emerged a single peak under the success condition and it was decreased by 89.09 and 55.56% compared with the second peak under the case no spray. Meanwhile, the *(dp/dt)* history at the upper end appeared a larger negative value. And its absolute value was decreased constantly with the increase of spray time. The negative value disappeared as the synergy effect was successful. This fully indicates that the reaction rate of syngas explosion was reduced by spray mist, and the synergy effect continued to enhance as the spray amount increased.

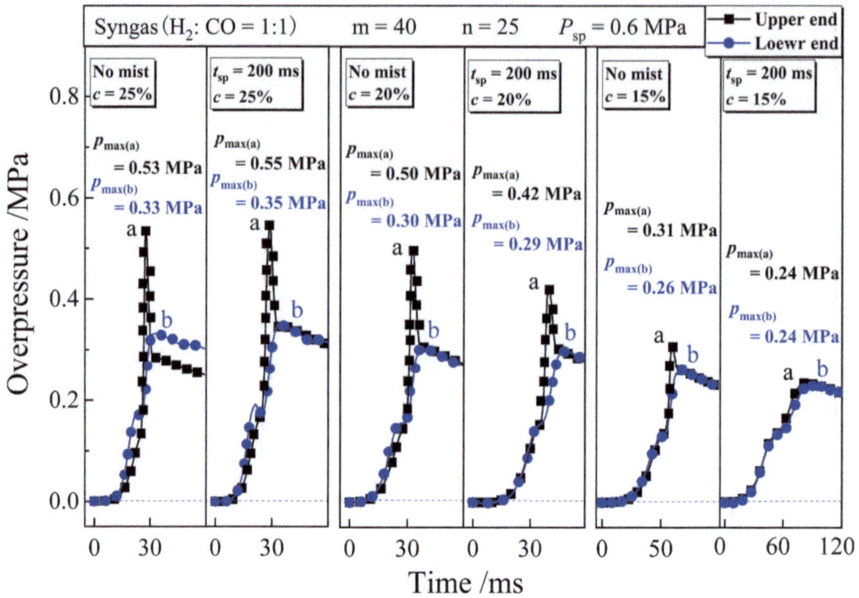

Fig. 5.8 Comparisons of pressure histories under different syngas concentrations (t_{sp} = 200 ms, P_{sp} = 0.60 MPa)

Table 5.2 p_{max} corresponding to different spray times and syngas concentrations

p_{max}/MPa	Below wire mesh			Above wire mesh		
	Syngas concentration					
t_{sp}/ms	c = 15%	c = 20%	c = 25%	c = 15%	c = 20%	c = 25%
0	0.24	0.30	0.33	0.31	0.50	0.53
200	0.24	0.29	0.35	0.26	0.42	0.55
300	0.13	0.28	0.33	0.14	0.40	0.51
400	—	0.28	0.32	—	0.38	0.47
500	—	0.15	0.29	—	0.14	0.41
600	—	—	0.18	—	—	0.17

Simultaneously, the concentration of syngas also had a significant impact on the synergy effect (t_{sp} = 200 ms, P_{sp} = 0.60 MPa), as shown in Fig. 5.10. Compared with 25% syngas concentration, the *(dp/dt)* histories at two ends were evidently decreased and became more significant as the syngas concentration decreased. Especially for the second peak value in *(dp/dt)* history at the upper end. And the negative value of history also showed a decreasing trend. Table 5.3 illustrates the effect of spray time on the $(dp/dt)_{max}$ at the upper and lower ends under different syngas concentrations. It clear explains that the explosion reaction rate was greatly reduced with the decrease of syngas concentration and the increase of spray time.

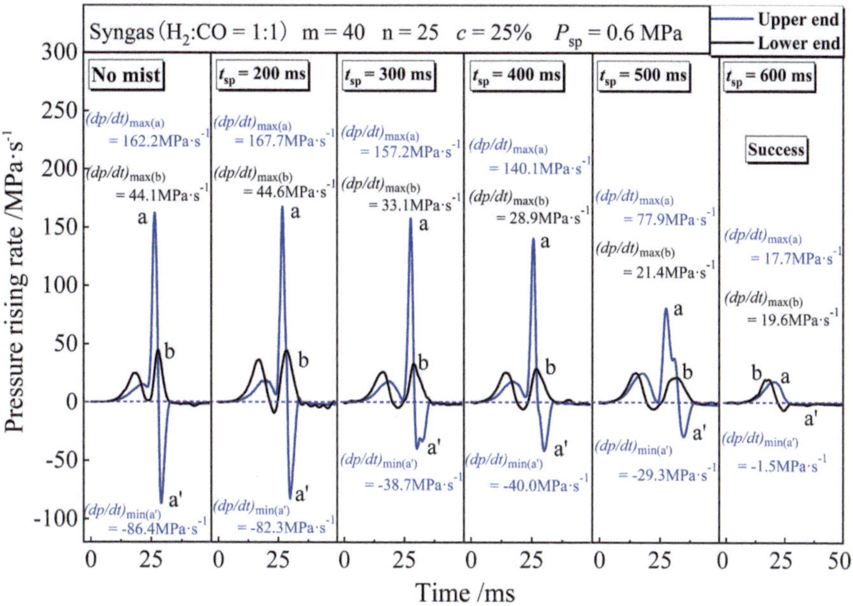

Fig. 5.9 Comparisons of *(dp/dt)* histories under different spray times ($c = 25\%$, $P_{sp} = 0.60$ MPa)

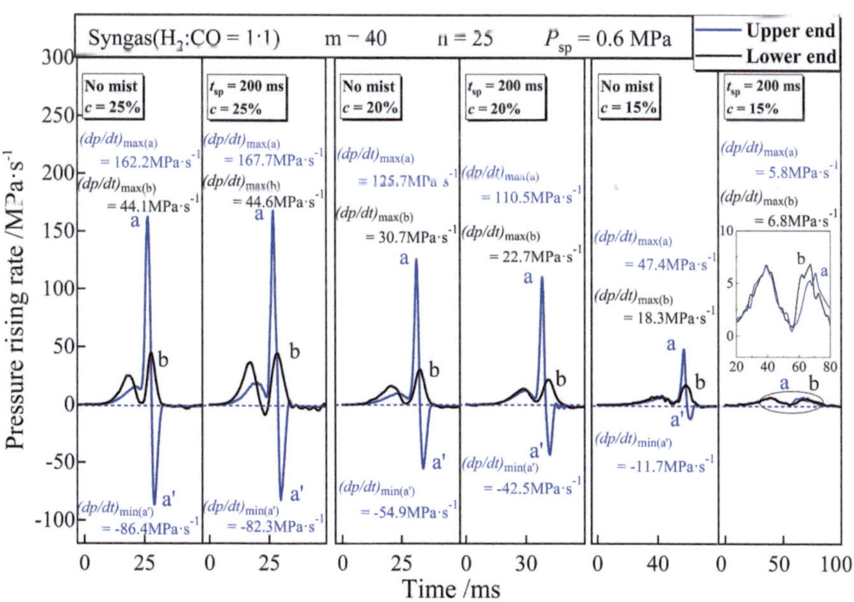

Fig. 5.10 Comparisons of *(dp/dt)* histories under different syngas concentrations ($t_{sp} = 200$ ms, $P_{sp} = 0.60$ MPa)

Table 5.3 $(dp/dt)_{max}$ corresponding to different spray times and syngas concentrations

$(dp/dt)_{max}$/MPa s^{-1}	Below wire mesh			Above wire mesh		
	Syngas concentration					
t_{sp}/ms	$c = 15\%$	$c = 20\%$	$c = 25\%$	$c = 15\%$	$c = 20\%$	$c = 25\%$
0	18.3	30.7	44.1	47.4	125.7	162.2
200	6.8	22.7	44.6	5.8	110.5	167.7
300	5.6	19.9	33.1	5.5	93.2	157.2
400	—	18.3	28.9	—	90	140.1
500	—	11.6	21.4	—	10.8	77.9
600	—	—	19.6	—	—	17.7

5.2.5 Analysis of Flame Temperature

Figure 5.11 illustrates the effect of spray time and syngas concentration on the flame temperature at the both ends under synergy inhibition. The upper part is the influence of spray time on flame temperature ($c = 25\%$, $P_{sp} = 0.60$ MPa), as can be seen in Fig. 5.11a. As the inhibition failed, the temperature history at the upper end was evidently higher than that of lower end. Especially for T_{max}. The change in flame temperature was corresponding to the change in pressure. As $t_{sp} = 200$ ms, the temperature history at the upper end was increased slightly, but it was evidently decreased with the increase of spray time. As $t_{sp} = 500$ ms, T_{max} at the upper and lower ends were decreased by 24.83% and 21.01% respectively. Especially the synergy inhibition was successful as $t_{sp} = 600$ ms. At this moment, the flame did not appear at the upper end and the temperature history also did not present any upward change. It indicates that the mist had an evident endothermic effect on the flame. Meanwhile, the heat generated by the explosion reaction did not dissipate in a timely manner due to the sealing effect of flame resistance system, leading to a slight increase in flame temperature at the lower end ($T_{max} = 1555$ °C).

Figure 5.11b presents the effect of syngas concentration on the temperature histories at both ends under synergy inhibition ($t_{sp} = 200$ ms, $P_{sp} = 0.60$ MPa). The concentration of syngas had an obvious impact on the flame temperature at the both ends of wire mesh. As the syngas concentration reduced, the temperature histories at the two ends also presented a decreasing trend. Meanwhile, the temperature history at the upper end was higher compared with the lower end. It was noticeable that it was detected by the thermocouple although the high-temperature gas passing through the flame resistance system did not ignite the syngas as the synergy effect was successful, further causing a certain degree of increase in the temperature history ($T_{max} = 234$ °C). This fully explains that the synergy inhibition of wire mesh and spray could greatly absorb the heat of explosion reaction and weaken its intensity. And the synergy effect became more significant with the increase of spray amount and the decrease of syngas concentration [5].

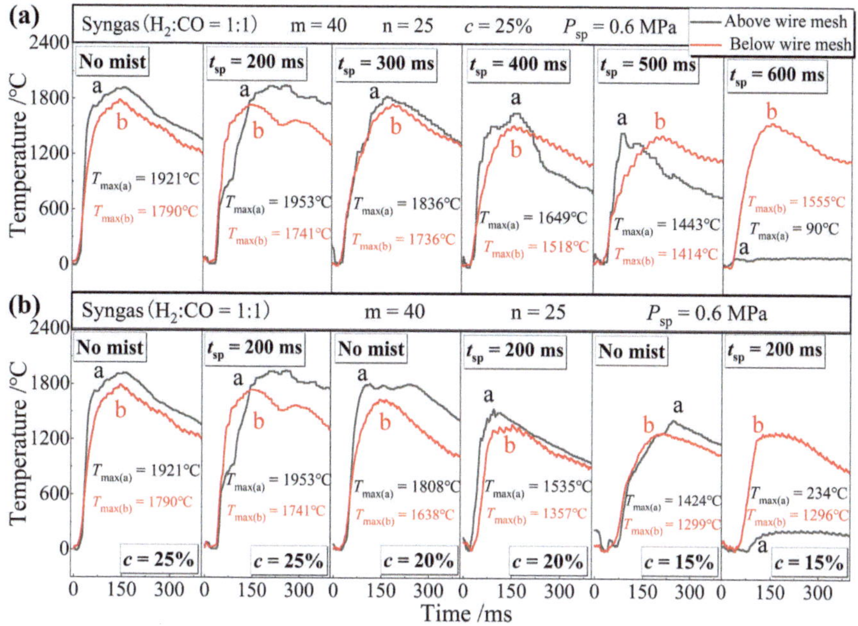

Fig. 5.11 Effects of spray time and syngas concentration on the flame temperature at the both ends under synergy inhibition

5.2.6 *Synergy Inhibition Mechanism*

To describe the synergy effect mechanisms of explosion resistance/inhibition, Fig. 5.12 presents the synergy inhibition processes of two protection methods on syngas explosion flame. The inhibitions of wire mesh and spray was attributed to the combined physical and chemical effects. After the premixed gas at the bottom end was ignited, the syngas underwent a rapid explosion reaction ($2OH = O + H_2O$, $OH + H_2 = H + H_2O$, $H + O_2 = OH + O$, $CO + OH = CO_2 + H$, $O + H_2O = OH + OH$, $CO + O_2 = CO_2 + O$, $H + O_2 = O + OH$) [6] and accelerated upward propagation with the two waves and three zone structures. At this moment, the flame temperature at the lower end was increased rapidly. Due to the obstruction of flame resistance system, the reflected pressure wave would disturb the flame front and accelerate its propagation. Thereafter, the continuous flame was dispersed into a large number of discrete structures after it entered the wire mesh, leading to the continuous increases in the contact area and heat absorption effect between the flame and wire mesh.

Meanwhile, the wire mesh presented an obvious wall effect on the explosion reaction process of syngas, further greatly weakening the number and reaction activation energy of free radicals during the explosion reaction [7]. Affected by two weakening effects, there were two results of flame quenching inside the wire mesh and the protecting failure. After passing through the wire mesh and reaching the ignition temperature, the syngas would be ignited and its flame would accelerate upward

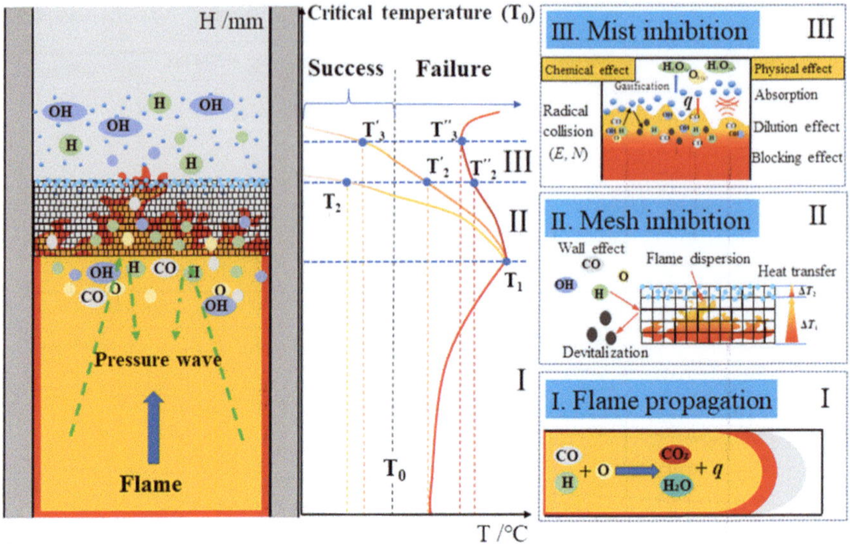

Fig. 5.12 Schematic diagram of synergy inhibition mechanism

propagation. As the mist existed, it exhibited a strong physical (the heat absorption and dilution) and chemical effects [8]. The flame temperature could be greatly reduced and reached below the critical temperature (the ignition temperature) as it passed through the wire mesh. Besides, the reaction rate would be slowed down and the water molecules as the third element would participate in the reaction process as the spray amount increased, leading to the obvious reductions in the number and energy of free radicals used to activate participation in the explosion reactions ($O + H_2 = H + OH$, $H + O_2 = O + OH$, $OH + H_2 = H + H_2O$). This further weakened the intensity of explosion reaction and the inhibition effect continued to increase as the spray amount increased.

5.3 Conclusions

The flame propagation characteristics of syngas explosion under the explosion resistance/inhibition synergy method were studied experimentally. The corresponding relationship between explosion parameters at the upper and lower ends under the success and failure conditions was analyzed. The change regular of explosion parameter with the changes of spray amount and syngas concentration was obtained and its synergy inhibition mechanism was revealed. The conclusion was as follows:

(1) Compared with the inhibition of single wire mesh, the weakening effect of gas explosion could be greatly enhanced after adding spray. However, the pressure history only presented an acceleration rise although the flame propagation emerged two acceleration processes under the success condition. Although

the flame propagation velocity appeared the slight increase, (dp/dt) and T_{max} presented an obvious decrease trend. And the velocity of flame propagation at the upper end was faster compared with the lower end as the inhibition failed. And the pressure change was also more significant. Nevertheless, the pressure at the upper end was slightly decreased than the lower end under the success condition. Besides, the disturbance of flow field near the flame resistance system at the lower end was enhanced with the increase of spray time, resulting in a slight increase in the velocity history. But the velocity history at the upper end showed a change trend of increasing firstly and then decreasing as the spray amount increased. And the stagnation time of flame inside the wire mesh was also increased. Additionally, the synergy effect became more significant and the stagnation time was longer as the syngas concentration decreased.

(2) The intensity of syngas explosion could be greatly weakened by synergy effect. As the spray amount increased, the pressures at the two ends were firstly increased and then decreased. It was mainly affected by the combined effect of the turbulence disturbance of wire mesh and heat absorption of mist on the flame. Meanwhile, the pressure at the upper end was evident higher than the lower end as the flame resistance failed, and its reduction degree was more significant compared with the lower end. As the inhibition was successful, the pressures at both ends were decreased. Under lean-burn condition, the pressure also presented a decreasing trend as the syngas concentration reduced. Besides, the explosion intensity and its reaction rate were also constantly decreased with the increase of spray amount.

(3) Under the synergy effect, the flame temperature was affected by the concentrations of spray and syngas, and its variation regular was similar as the pressure. As the synergy inhibition failed, the flame temperature at the upper end was evident higher than the lower end. And the temperature history showed a change trend of firstly increasing and then decreasing as the spray amount increased. Meanwhile, the flame temperature was reduced sequentially as the decrease of syngas concentration. Especially the flame temperature at the upper end did not presented an obvious increase as the inhibition was successful, but its history still showed varying degrees of increase under the action of high-temperature gas.

For practical engineering applications, the effective inhibition method should be adopted. Three suggestions were proposed based on the research conclusions: (1) The disturbances of wire mesh and spray mist could improve the propagation velocity of flame, so the amount of mist sprayed should ensure that the endothermic effect was higher than the enhancement effect of turbulence disturbance. (2) To improve the inhibition effect of spray, additives that could enhance the physical and chemical effects should be selected and dissolved in the water mist. (3) To improve the flame inhibition effect, the alloy materials with a high thermal conductivity and quenching performance of flame resistance system should be selected preferentially.

References

1. Cao XY, Zhou X, Wang Z, Zhou JY, Wu ZR, Lu YW, Lin CD. Study on the synergy inhibition of ultrafine water mist and metal wire mesh on the syngas explosion. Int J Hydrogen Energy. 2024;70:1089–100.
2. Cao X, Wang Z, Wang Z, Lu Y, Sun S, Xu J. Experimental research on the flame resistance characteristics of wire mesh for syngas explosion. Process Saf Environ Protect. 2023;175:34–47.
3. Wang S, Xiao G, Wang W, Mi H. Study of the process of deflagration flame behavior in disturbance conditions. Exp Therm Fluid Sci. 2022;139:110710.
4. Jin K, Wang Q, Duan Q, Chen J, Sun J. Effect of metal wire mesh on premixed H_2/air flame quenching behaviors in a closed tube. Process Saf Environ Protect. 2021;146:770–8.
5. Cao X, Ren J, Bi M, Zhou Y, Li Y. Experimental research on the characteristics of methane/air explosion affected by ultrafine water mist. J Hazard Mater. 2017;324:489–97.
6. Davis S, Joshi A, Wang H, Egolfopoulos F. An optimized kinetic model of H_2/CO combustion. Proc Combust Inst. 2005;30(1):1283–92.
7. Saiki Y, Fan Y, Suzuki Y. Radical quenching of metal wall surface in a methane-air premixed flame. Combust Flame. 2015;162(10):4036–45.
8. Jiang H, Bi M, Huang L, Zhou Y, Gao W. Suppression mechanism of ultrafine water mist containing phosphorus compounds in methane/coal dust explosions. Energy. 2022;239:121987.

Chapter 6
Research on the Synergy Effect of Resistance/Inhibition on the Syngas Explosion

6.1 Experimental Apparatus and Procedures

Figure 6.1 illuminates the experimental setup for the synergistic effect of resistance/inhibition on the syngas explosion. The main structure has been mentioned in the previous experiments. The pressure spray system and wire mesh explosion-resistant system were added to the experimental apparatus. The pressure spray system was composed of fine spray nozzle, one-way valve, solenoid valve and water storage tank. The nozzle was installed on the lower end of top flange through threaded connection method. The water storage tank was filled with the deionized water to eliminate the factors affecting explosion inhibition and the spray pressure was provided by the gas cylinders. The mist parameter was altered by changing the nozzle type and spray pressure. The moment and duration of spray were achieved by controlling the opening and closing of solenoid valve, further changing spray amount at the upper end. Under the action of pressure, the fine atomization solid nozzle conducted spray at an angle of 60°. The cross section of the vessel was 150 mm × 185 mm. The internal high of vessel was 910 mm and the height of the nozzle was 40 mm after connecting to the top flange. By calculation, the distance between the nozzle and the wire mesh explosion-resistant system was 210 mm. At a spray angle of 60°, the corresponding mist zone width at a height of 210 mm was 144.6 mm and it was less than the wall surface distance. This indicates that the water mist produced by pressure spray was mainly distributed inside the vessel. However, the movement velocity of micrometer scale mist was sharply reduced as the distance increased due to the drag force although the mist near the nozzle had a high movement velocity. And most of the mist would be suspended in the upper end of the flame resistance system within the millisecond spray and standing times.

The wire mesh explosion resistance system [effective area (A) = 100 mm × 100 mm] was composed of multi-layer stainless steel wire mesh, the upper and lower pressing plates [thickness (δ) = 3 mm] and four support rods. Stainless steel with a thermal conductivity of 21.5 W m^{-1} K^{-1} had a good corrosion resistance and

Z. Wang and X. Cao, *Flame Resistance Characteristics and Its Influence Mechanism*,
https://doi.org/10.1007/978-981-95-6119-3_6

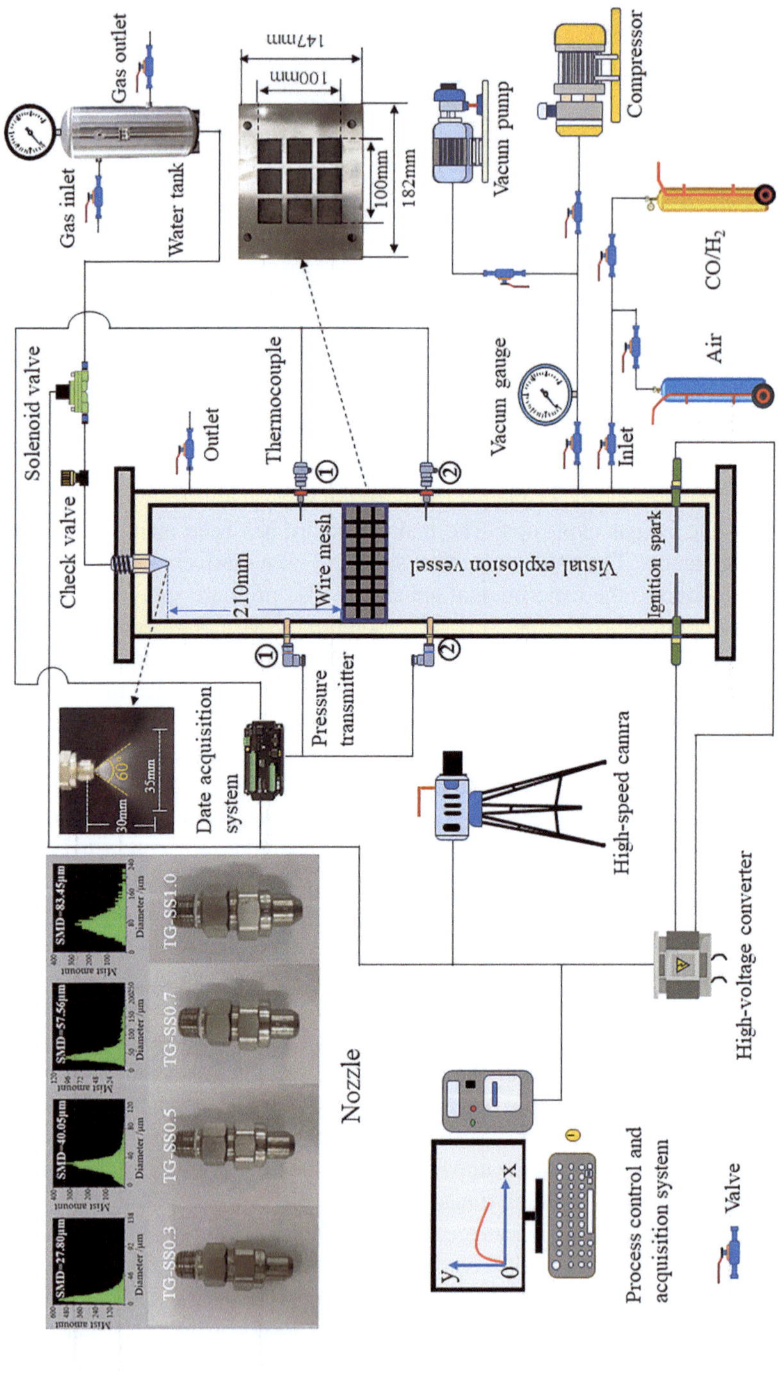

Fig. 6.1 Schematic diagram of experimental apparatus structure

strong tensile strength. The wire mesh was compressed by two pressing plates and pre tightening force was provided by bolts. And the explosion resistance system was fixed to the support rod and the height was determined to ensure that the flame fully developed (H = 650 mm). Considering the gap between the explosion resistance device and the vessel wall surface, the structure size of press plates was designed and processed based on the cross-sectional area of explosion vessel to achieve close integration. Meanwhile, the gap between the wall surface and the press plate was blocked by the vacuum sealing mud [1].

Syngas and air were mixed inside the vessel. The high-voltage discharge method with a discharge duration of 100 ms was used to ignite premixed gas (U = 8 kV). And a pair of ignition electrodes with a gap of 6 mm was set 110 mm apart from the bottom of the vessel to activate the explosion. A high-speed camera with an acquisition frequency of 4000 fps was adopted to record the flame propagation characteristics of syngas explosion under the synergistic effect of resistance/inhibition. The accurately flame position was determined using the pixel coordinates of post-processing system. And its change at an interval of 0.25 ms was determined and the propagation velocity of flame inside the vessel was calculated through proportional conversion. Two high-frequency thermocouples [Acquisition range (A) = 2000 °C] and two high-frequency pressure transmitters [Acquisition frequency (f) = 200 kHz] were installed at a distance of 50 mm from the upper and lower ends of explosion resistance system to record the pressure and temperature changes. To ensure their corresponding accuracy in terms of time, the spraying system, ignition system, pressure acquisition system, temperature acquisition system and image acquisition system were controlled by a program control and data acquisition system during the experiment (200 kHz/per channel). And their order and interval time were determined.

The mist was produced by pressure spray during the experiment. Four types of solid conical nozzles (Nozzle type: TG-SS 0.3, 0.5, 0.7, and 1.0) produced by spraying system Co., Ltd were selected. The number represented the size of the nozzle. The increase of nozzle type meant the increase in nozzle size, further changing the mist parameter (see in Table 6.1). The mist diameter and velocity below 30 mm were measured under four nozzle types at 0.60 MPa and five spray pressures (P_{sp} = 0.20, 0.40, 0.60, 0.80, and 1.00 MPa) through the phase doppler particle analyzer. The spray rate was determined by the amount of water mist collected per unit time. Mist parameters under different working conditions were shown in Table 6.1. The premixed gas was directly configured inside the vessel [(Component ratio) H_2: CO = 1:1; (Syngas concentration) c = 25%]. Before the experiment, the explosion-resistant system was installed inside the vessel [mesh number (m) = 40; (layer number) n = 25]. The initial pressure at the ignition was 0.10 MPa. The spray time (t_{sp}) was 400 ms under four spray nozzle types and the spray time was 600 ms under five spray pressures. And the mist suspended inside the vessel for 100 ms after spray. Afterwards, the premixed gas was ignited. After ignition, the flame propagated from the bottom to top end of the vessel under the synergistic effect of explosion resistant/inhibition systems. Through repetitive experiments, the error values of $\triangle P_{max}$, v_{max} and T_{max} were 5.4%, 4.6% and 9.1%, respectively.

Table 6.1 Mist parameters under different nozzle types and spray pressures

P_{sp}/MPa	0.20	0.40	0.60				0.80	1.00
Nozzle/TG-SS	0.5	0.5	0.3	0.5	0.7	1.0	0.5	0.5
d_{32}/μm	75.60	52.38	27.80	40.05	57.56	83.45	34.30	26.98
Q_{mist}/g s^{-1}	3.35	4.45	3.21	5.42	8.83	40.25	6.26	6.97
v_{mist}/m s^{-1}	7.03	10.50	9.20	12.92	14.59	17.33	13.51	14.76

6.2 Results and Discussions

6.2.1 Analysis of Explosion Resistance/inhibition Characteristics

Figure 6.2 presents the correspondence between the explosion parameters and flame propagation of synthetic gas as the synergistic effect failed ($P_{sp} = 0.60$ MPa; $t_{sp} =$ 400 ms; Nozzle TG-SS 0.3). Under this working condition, the flame passed through the wire mesh and ignited the premixed gas at the upper end. A clear correspondence between flame propagation and pressure rise was presented. After the premixed gas was ignited, the flame developed in a "spherical" structure. Due to the weak flame during the initial stage, the pressure history did not present an obvious upward trend. Afterwards, the flame transformed to a "finger" structure and accelerated upward propagation under the constraint of wall surface. At 7.25 ms, the pressure history at the lower end began to rise. As the flame accelerated, the rise rate of pressure was evidently increased. However, there was a slight delay in the pressure rise at the upper end (t = 9.25 ms). At 13.25 ms, the velocity history appeared the first peak [$v_{(1)} =$ 43.3 m/s]. As the curvature radius of flame front raised, its propagation velocity was slowed down. As approached the wire mesh, the flame appeared the second peak at 16.50 ms [$v_{(2)} = 42.2$ m/s]. Afterwards, the flame entered the wire mesh and (dP/dt) history showed the first peak at 17.50 ms [$(dP/dt)_{max(1)} = 25.8$ MPa s^{-1}]. The flame passed through the wire mesh after stopped inside the wire mesh for 5.50 ms and ignited the premixed gas at the upper end (t = 23.00 ms). And the (dP/dt) history was obviously decreased. After passing through the wire mesh, the flame accelerated propagation and the velocity history at the under end reached its peak value (144.1 m/s) at t = 25.50 ms. Due to the inhibition failure, the pressure history at the lower end presented the second upward process under the action of pressure wave reflected from the top end of the vessel [$(dP/dt)_{max(2)} = 28.9$ MPa s^{-1}]. Afterwards, the rate of pressure rise was decreased as the lower flame extinguished. ΔP_{max} was 0.32 MPa at 39.25 ms. Compared with the lower end, the pressure history at the upper end was increased evidently. The temperature at the upper end was also increased ($T_{max} =$ 1649 °C). Simultaneously, the (dP/dt) history at the upper end reached its peak value at 28.25 ms. The reflected pressure wave generated at the upper end of vessel would pass through the wire mesh, causing a sharp decrease in the rise rate of pressure at the upper end. The (dP/dt) history even showed an obvious negative value [$(dP/dt)_{min}$ = -40.0 MPa s^{-1}]. At 30.00 ms, ΔP_{max} was 0.47 MPa. Afterwards, the pressure was decreased evidently as the flame extinguished.

Figure 6.3 presents the correspondence between the explosion parameters and flame propagation as the synergistic effect was successful ($P_{sp} = 0.60$ MPa; $t_{sp} =$ 400 ms; Nozzle TG-SS 0.7). Under this working condition, the flame did not pass through the wire mesh to ignite the premixed gas of upper end. The acceleration propagation of flame resulted in a sharp increase in pressure. There was still a clear correspondence between flame propagation and pressure rise. However, the moments of pressure rise at both ends were slightly earlier compared with the failure (t =

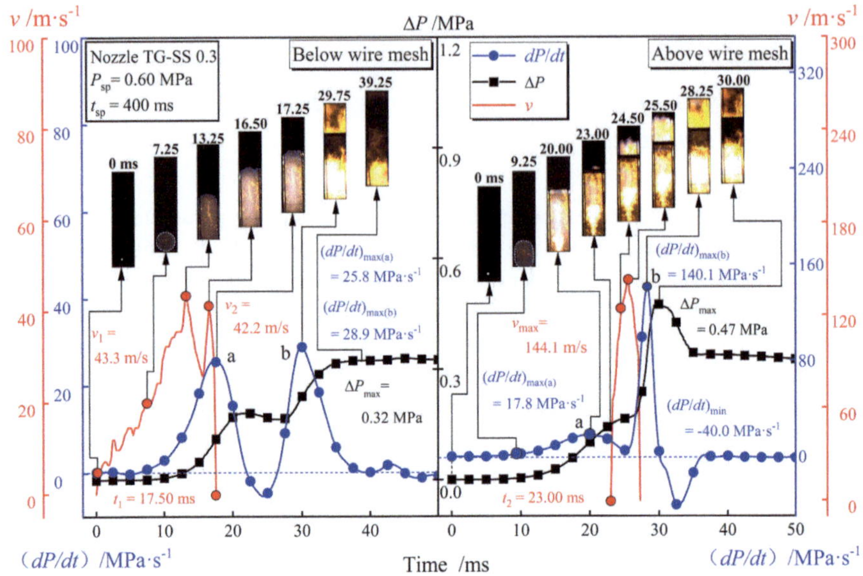

Fig. 6.2 Correspondence of explosion parameters under explosion resistance failure (P_{sp} = 0.60 MPa; t_{sp} = 400 ms; Nozzle TG-SS 0.3)

5.50 and 6.00 ms). Under the inhibition success, the flame propagation exhibited two acceleration processes and the velocity history appeared two peak values. The velocity peak appeared was slightly earlier, but the its change was not significant [$v_{(1)}$ = 43.4 m/s; $v_{(2)}$ = 41.6 m/s]. Nevertheless, the pressure of lower end only showed an acceleration process, and the peak value was evidently decreased. After inhibiting success, the explosion intensity was decreased. ΔP_{max} of two ends were decreased from 0.47 MPa and 0.32 MPa to 0.16 MPa and 0.17 MPa, respectively. And $(dP/dt)_{max}$ were decreased from 140.1 MPa s^{-1} and 28.9 MPa s^{-1} to 17.2 MPa s^{-1} and 22.4 MPa s^{-1}, respectively. It clear that the pressure drop of upper end was more significant. Compared with the inhibition failure, the temperature of lower end was slightly increased due to the strong sealing effect produced by the wire mesh (T_{max} = 1522 °C). However, the temperature of upper end (T_{max} = 116 °C) was evidently decreased and did not reach the ignition temperature of syngas. After entering wire mesh, the flame stopped and reached its brightest extent at 20.00 ms. Afterwards, the flame accelerated extinguishing as the explosion reaction progressed.

6.2.2 *Analysis of Flame Propagation Processes*

Figure 6.4 shows the effect of nozzle type on the flame propagation process under synergistic effect (P_{sp} = 0.60 MPa; t_{sp} = 400 ms). The flame accelerated propagation

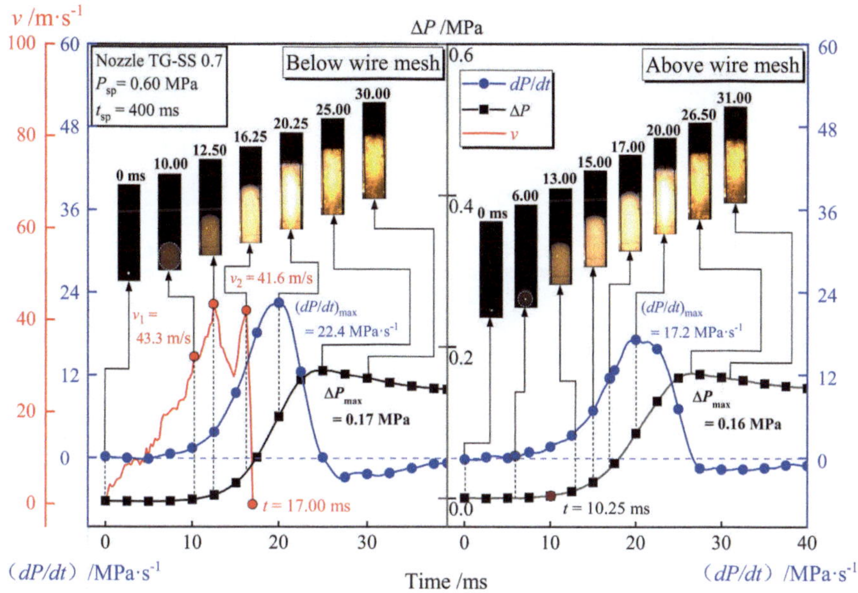

Fig. 6.3 Correspondence of explosion parameters under explosion resistance success (P_{sp} = 0.60 MPa; t_{sp} = 400 ms; Nozzle TG-SS 0.7)

in the early stages and underwent a transition from a "spherical" to a "finger" structure. Based on the velocity history in Fig. 6.5, the initial velocity history was slightly increased as the nozzle type increased. This was because that the change in nozzle type resulted in the changes of mist parameters. The larger mist amount and diameter would disturb the flow field at the lower end of wire mesh and increase the emission effect of wire mesh on the pressure wave, thereby causing a certain disturbance effect on the initial flame front. Under no spray and Nozzle TG-SS 0.3 and 0.5 nozzles conditions, the synergistic inhibition failed. At this moment, the flame underwent three acceleration processes (I, II and III stages). Namely, self-acceleration, mesh acceleration, and turbulent acceleration processes.

As the nozzle type increased, the two peaks of velocity history at the lower end were slightly increased. Although the flame front of upper end was disturbed and enhanced by the actions of wire mesh and mist, the velocity history showed an evident decrease trend as the mist concentration increased (from 153.7 to 84.1 m/s). This indicates that the inhibition effect of mist concentration increase was higher than the enhancement effect of turbulence generated by the wire mesh and mist under the pressure spray. Meanwhile, the dwell time of flame inside the wire mesh was increased (from 2.50 to 11.00 ms). Especially for the Nozzle TG-SS 0.7 and 1.0 nozzles, the synergistic inhibition was successful. At this moment, the flame only underwent two acceleration processes (I and II stages).

Figure 6.6 shows the effect of spray pressure on the flame propagation process of syngas explosion (Nozzle TG-SS 0.3; t_{sp} = 600 ms). Combined with Fig. 6.7,

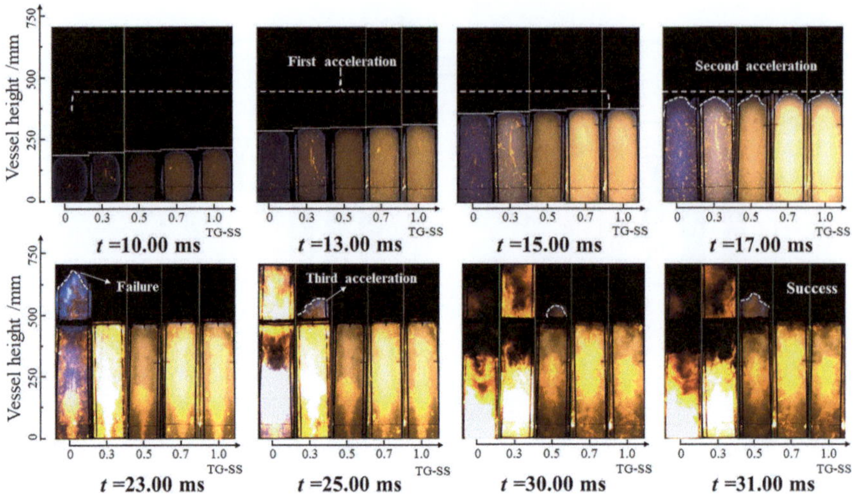

Fig. 6.4 Effect of nozzle type on the flame propagation process under synergistic inhibition (P_{sp} = 0.60 MPa; t_{sp} = 400 ms)

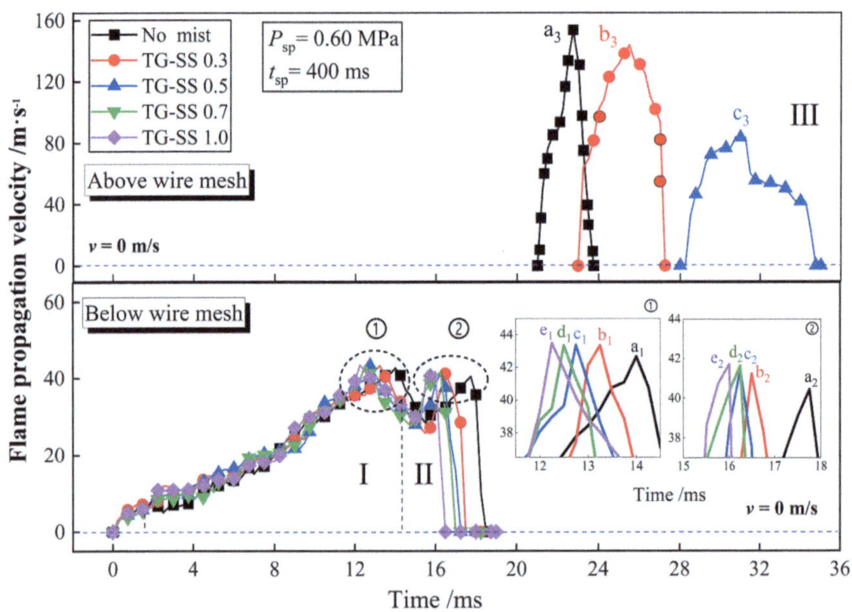

Fig. 6.5 Effect of nozzle type on the velocity history under synergistic inhibition (P_{sp} = 0.60 MPa; t_{sp} = 400 ms)

the flame could pass through the wire mesh and the flame resistance failed under the condition without spray. With the increase of spray pressure, the synergetic inhibition showed a change trend of failure—success—failure. In the early stage, the flame propagation at the lower end showed the first acceleration process, and the velocity history was slightly increased with the enhancement of spray pressure. This was because that the increase of spray pressure resulted in the enhancement of flow field disturbance at the lower end. As it approached the wire mesh, the flame contracted toward the central area and accelerated propagation. The velocity history appeared two peaks and presented a slightly increasing change trend. In the absence of spray, the flame first passed through the wire mesh and ignited the premixed gas at the upper end. With the increase of spray pressure (0 MPa $\leq P_{sp} \leq$ 0.40 MPa), the time for the flame to pass through the wire mesh was increased in sequence. As P_{sp} = 0.60 MPa, the flame failed to pass through the wire mesh. With the increase of spray pressure (0.80 MPa $\leq P_{sp} \leq$ 1.00 MPa), the synergetic inhibition failed and the time of flame through wire mesh to ignite premixed gas was advanced successively. In addition, the velocity history after passing through the wire mesh showed a trend of first decreasing and then increasing with the increase of spray pressure. The dwell time of flame inside the wire mesh presented a change trend of increasing first and then decreasing. This was mainly because that the increase of spray pressure changed the mist parameters, further leading to different inhibition results due to the combined effect of mist parameters [2].

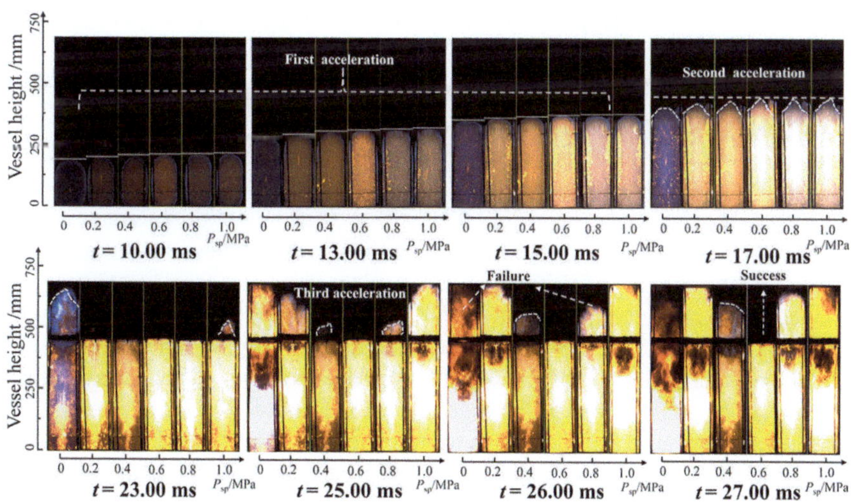

Fig. 6.6 Effect of spray pressure on the flame propagation process under synergistic inhibition (Nozzle TG-SS 0.3; t_{sp} = 600 ms)

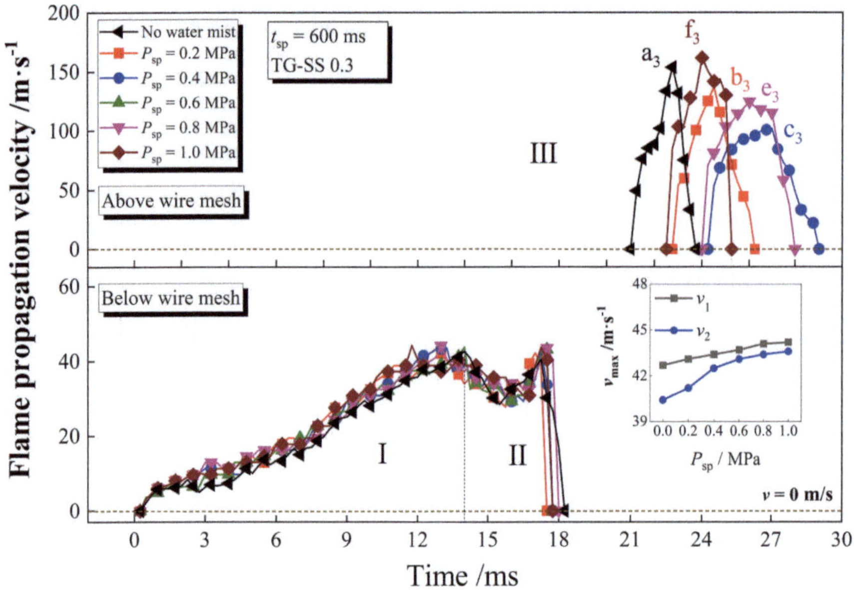

Fig. 6.7 Effect of spray pressure on the velocity history under synergistic inhibition (Nozzle TG-SS 0.3; $t_{sp} = 600$ ms)

6.2.3 Analysis of Explosion Resistance/inhibition Pressure

Figure 6.8 shows the effect of nozzle type on the pressure under the synergistic inhibition ($P_{sp} = 0.60$ MPa; $t_{sp} = 400$ ms). As the nozzle type changed, the pressure history presented the same trend as the velocity history. Under no spray, the explosion resistance failed and the pressure at the upper end was greater than that of lower end. After spraying, the synergistic effect still showed inhibition failure at Nozzle TG-SS 0.3 and 0.5. However, the pressure at the upper end was reduced under the action of mist compared with the lower end. And the reduction degree was obviously enhanced as the nozzle type increased. ΔP_{max} was reduced from 0.53 to 0.38 MPa. Especially for Nozzle TG-SS 0.7 and 1.0, the synergistic inhibition was successful and the pressures at both ends showed a further decrease. ΔP_{max} was reduced from 0.38 to 0.16 MPa. However, the pressure histories at two ends were similar and the appearance moment of ΔP_{max} was almost the same as the synergistic inhibition was successful. This indicates that spray could effectively enhance the inhibition effect of wire mesh explosion-retardant, thereby reducing the explosion intensity.

Figure 6.9 presents the effect of spray pressure on the pressure history under synergistic inhibition (Nozzle TG-SS 0.3; $t_{sp} = 600$ ms). It clear that the pressures at both ends presented a change trend of decreasing first and then increasing with the increase of spray pressure. The corresponding synergistic effect exhibited an inhibition regular of "failure"—"success"—"failure". As the spray pressure increased (0 MPa $\leq P_{sp} \leq$ 0.40 MPa), the pressure at the lower end did not appeared the obvious

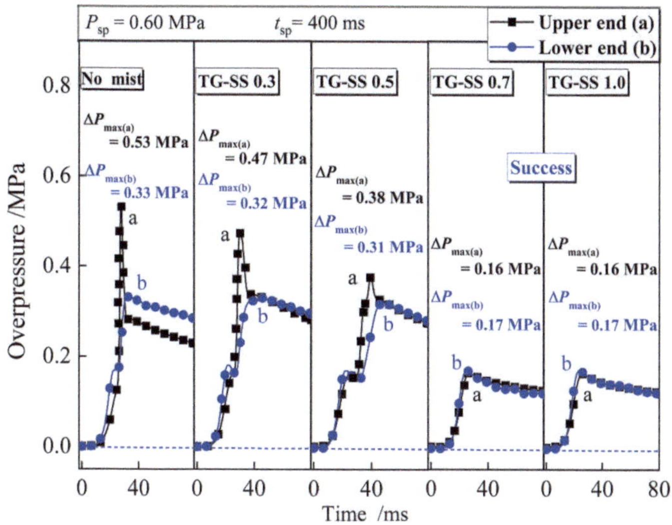

Fig. 6.8 Effect of nozzle type on the pressure history under synergistic inhibition ($P_{sp} = 0.60$ MPa; $t_{sp} = 400$ ms)

change. But the pressure at the upper end showed an evident decrease trend. ΔP_{max} was reduced from 0.53 to 0.41 MPa. As $P_{sp} = 0.60$ MPa, the synergistic effect of wire mesh and spray could effectively inhibit the flame propagation of syngas explosion, resulting in an obvious reduction in pressure ($\Delta P_{max} = 0.17$ and 0.18 MPa). However, the synergistic effect failed again as the spray pressure increased (0.60 MPa $\leq P_{sp} \leq$ 1.00 MPa). The flame passed through the wire mesh and ignited the premixed gas at the upper end, leading to a significant increase in the pressures at both ends, even exceeding the pressure without spray. This was because that the increase of spray pressure changed the mist parameters. Although the spray amount was increased and mist diameter was decreased, the inhibition effect could be enhanced. However, the disturbance effect of mist on the flame was also enhanced. The acceleration propagation of flame resulted in a significant increase in the rate of pressure rise. The rate of heat accumulation inside the vessel was also significantly increased, resulting in the evident increase of pressure. Besides, the vapor pressure generated by mist gasification at high temperature would also evidently affect the pressure inside the vessel. This also fully indicates that synergistic inhibition was affected by the comprehensive effects of mist diameter, concentration and its motion velocity.

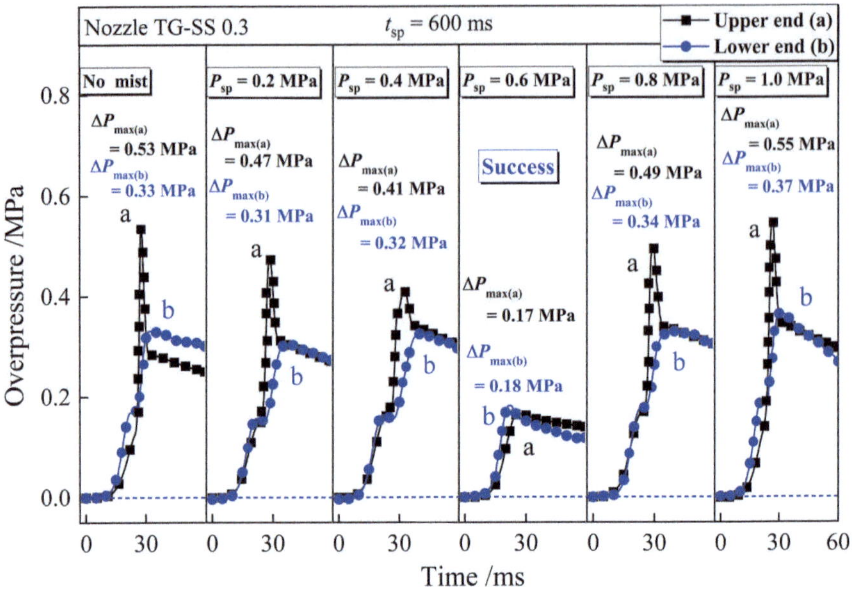

Fig. 6.9 Effect of spray pressure on the history under synergistic inhibition (P_{sp} = 0.60 MPa; t_{sp} = 600 ms)

6.2.4 Analysis of Explosion Resistance/inhibition Pressure Rise Rate

Figure 6.10 illuminates the effect of nozzle type on the (*dP*/*dt*) history under synergistic inhibition (P_{sp} = 0.60 MPa; t_{sp} = 400 ms). The variation regular of (*dP*/*dt*) history was similar to that of pressure history. As the synergistic inhibition failed, the (*dP*/*dt*) history exhibited two peaks and the second peak was evidently greater than the second peak. Compared with the lower end, the (*dP*/*dt*) history at the upper end was larger and its changes was more significantly, even showing larger negative value. After spraying, (*dP*/*dt*) was decreased. And the peak value of (*dP*/*dt*) was continuously decreased and the reduction extent of upper end was greater as the nozzle type increased. Especially for Noszle TG-SS 0.7 and 1.0, the synergistic inhibition was successful and (*dP*/*dt*) was further decreased. The (*dP*/*dt*) history only presented a single peak value and the peak at the lower end was slightly higher than the upper end.

Figure 6.11 presents the effect of spray pressure on (*dP*/*dt*) history (Nozzle TG-SS 0.3; t_{sp} = 600 ms). With the increase of spray pressure, the peak value of (*dP*/*dt*) history also showed a trend of first decreasing and then rising. The negative value of corresponding (*dP*/*dt*) history also exhibited the same variation regular. Especially the synergistic inhibition was successful and the (*dP*/*dt*) history only showed a single peak as P_{sp} = 0.60 MPa. The two peaks of (*dP*/*dt*) history at both ends was decreased to 17.7 MPa s^{-1} and 19.6 MPa s^{-1}, respectively. The rise rate

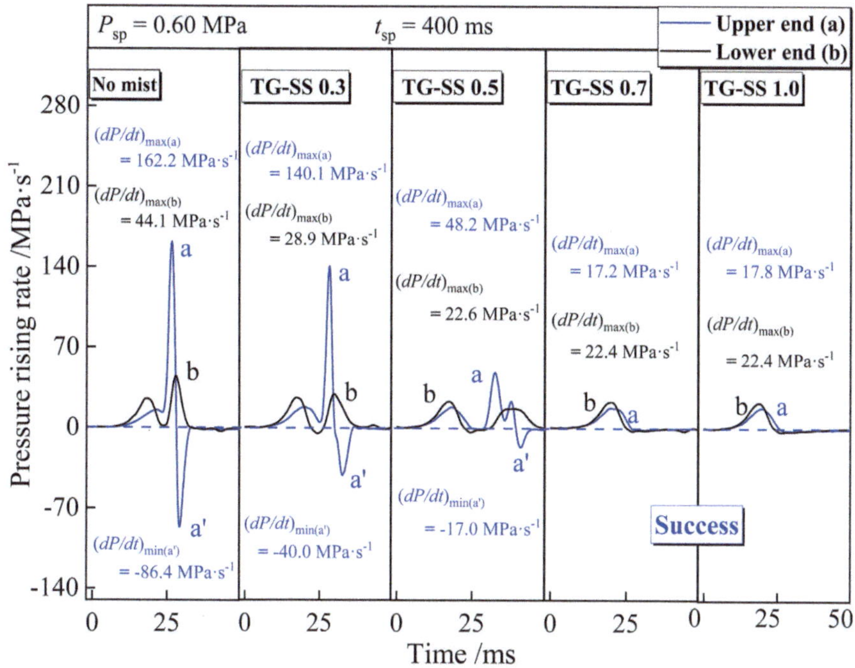

Fig. 6.10 Effect of nozzle type on the (dP/dt) history under synergistic inhibition ($P_{sp} = 0.60$ MPa; $t_{sp} = 400$ ms)

of pressure could directly reflect the weakening effect of synergistic inhibition on explosion reaction. It clear that the synergistic effect of resistance/inhibition could weaken the intensity of syngas explosion. Nevertheless, the mist parameter could greatly affect the weakening degree and even appear an enhancing effect.

6.2.5 Analysis of Flame Temperature

Figure 6.12 shows the effect of nozzle type on the temperature history under synergistic inhibition ($P_{sp} = 0.60$ MPa; $t_{sp} = 400$ ms). The variation regular of temperature history was similar to that of pressure history. The flame passed through the wire mesh under no spray, causing an obvious increase of temperature at the upper end and was higher than the lower end. After spraying, the temperature of upper end would also be higher than that of lower end as the synergistic inhibition failed. The mist parameters were changed with the variation of nozzle type. The comprehensive effect of changes in mist parameters leaded to an obvious decrease in flame temperature. T_{max} at the both ends were decreased from 1921 to 1505 °C and from 1790 to 1455 °C, respectively. As the nozzle type increased, the synergistic inhibition was successful. Compared with the failure, the decrease extent of T_{max} at the

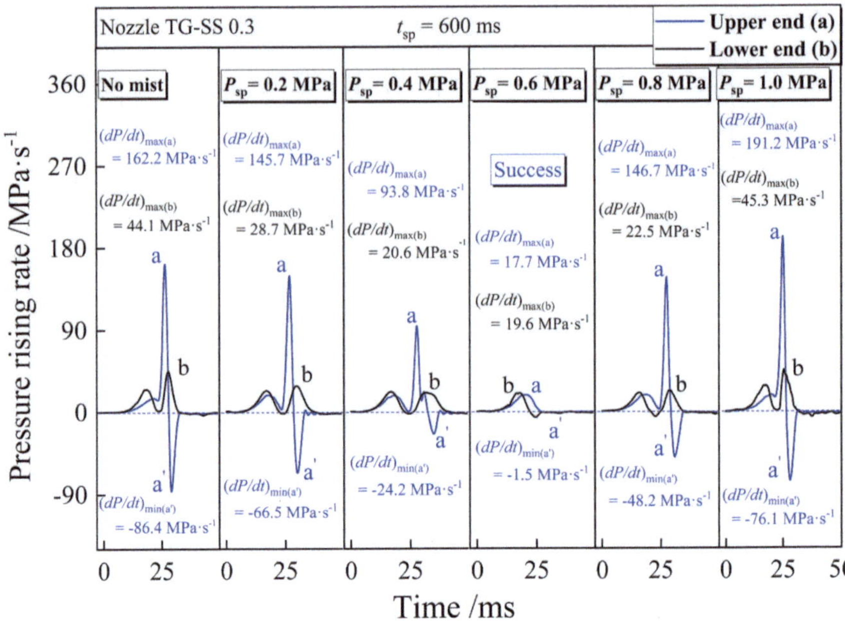

Fig. 6.11 Effect of spray pressure on the (dP/dt) history under synergistic inhibition (P_{sp} = 0.60 MPa; t_{sp} = 600 ms)

upper end was more evidently and the temperature history did not present an obvious upward trend. However, the wire mesh could effectively prevent the heat generated by the explosion from being released to the upper end as the explosion resistance was successful, resulting in the increase of heat accumulation at the lower end and a slight increase in temperature. Figure 6.13 presents the effect of spray pressure on the temperature history under synergistic inhibition (Nozzle TG-SS 0.3; t_{sp} = 600 ms). After spraying, the temperature history presented a significant downward trend. The increase of spray amount and the decrease of mist diameter leaded to the obvious decrease of flame temperature as the spray pressure increased. Especially, the flame of upper end could be completely inhibited as P_{sp} = 0.60 MPa. The mist exhibited a stronger heat absorption and dilution effects. Hence, the inhibition succeeded at a spray pressure of 0.60 MPa. However, the increase of spray pressure also leaded to the increases of spray amount and mist motion velocity, further resulting in the enhancement of disturbance effect on the flow field and flame front and causing the failure of synergy effect. This further leaded to the flame retardant failure and a continuous increase in temperatures at both ends of wire mesh. This also fully indicates that the addition of mist under synergistic inhibition could effectively reduce the temperature of explosion flame, but it was also evidently affected by the comprehensive effect of mist parameters.

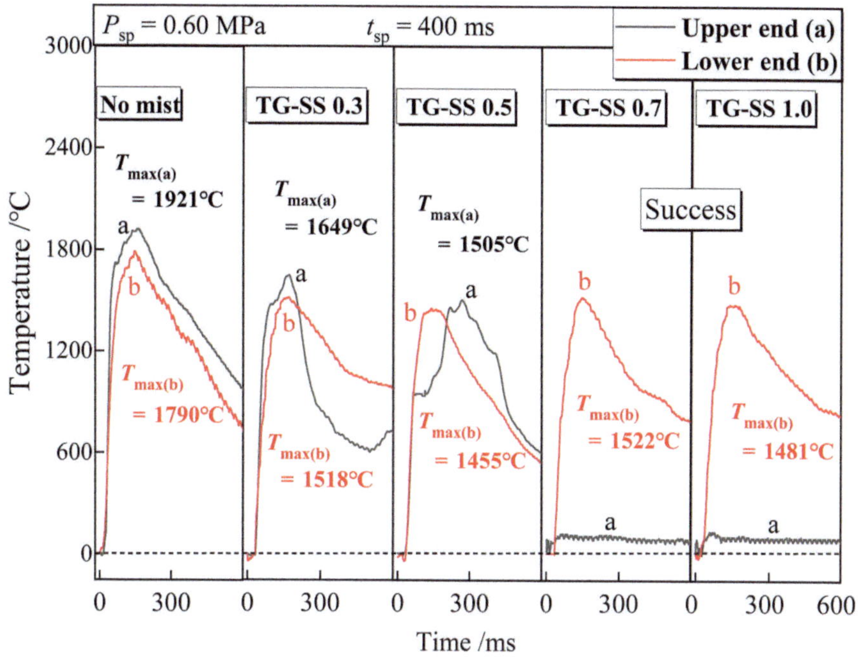

Fig. 6.12 Effect of nozzle type on temperature history under synergistic inhibition ($P_{sp} = 0.60$ MPa; $t_{sp} = 400$ ms)

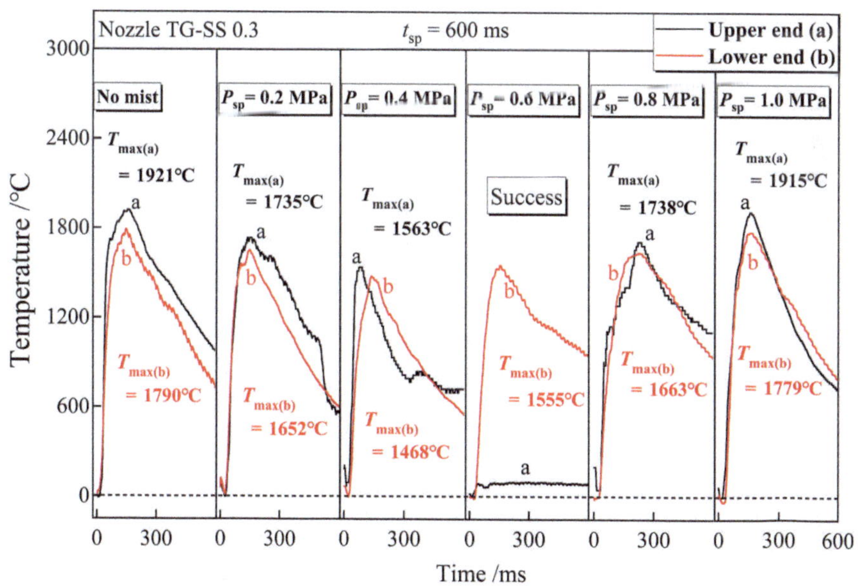

Fig. 6.13 Effect of spray pressure on temperature history under synergistic inhibition ($P_{sp} = 0.60$ MPa; $t_{sp} = 600$ ms)

6.2.6 Analysis of Explosion Resistance/inhibition Mechanism

Figure 6.14 presents the synergistic mechanism of resistance/inhibition on syngas explosion by wire mesh and water mist. The pressure wave struck the wire mesh before the flame and partial pressure wave would occur reflection. Nevertheless, the remaining pressure wave would be dispersed and scattered at different angles after passing through the wire mesh, resulting in a greatly weakening of explosion intensity [3]. Subsequently, the continuous flame entering the narrow channel of wire mesh was dispersed into a large number of small flame units. The increase of flame front leaded to an enhancement in the contact area between the wire mesh and flame, thereby promoting the enhancement of heat absorption capacity. Meanwhile, the syngas explosion was accompanied by a large number of reactive free radical chain reactions ($CO + O_2 = CO_2 + O$, $O + H_2O = OH + OH$, $CO + OH = CO_2 + H$, $H + O_2 = OH + O$, $H + O_2 = O + OH$, $OH + H_2 = H + H_2O$, $2OH = O + H_2O$) [4]. A large number of highly active free radicals were destroyed after contacting with the wire mesh, and the energy and collision probability were also sharply reduced, leading to a decrease in the flame propagation ability of syngas explosion. Under the combined effect of the above two factors, the explosion resistance was successful as the temperature at the upper interface of wire mesh did not reach the ignition temperature of synthetic gas. On the contrary, the explosion-retardant failed.

After adding mist, most of the water mist would disperse in the space of wire mesh upper end. However, part of mist would drip down onto the surface of wire mesh to form a water-containing mesh layer. In the lower part, the flame propagated in the space without water mist. After reaching water containing mesh layer, the flame would be synergistically inhibited by the wire mesh and water mist (in Fig. 6.14). Both the wire mesh and water mist all exhibit strong heat absorption inside the water-containing mesh layer. Meanwhile, the water vapor produced from mist gasification would existed inside the wire mesh, leading to a significant increase in flame propagation resistance. This further enhanced the physical and chemical interaction

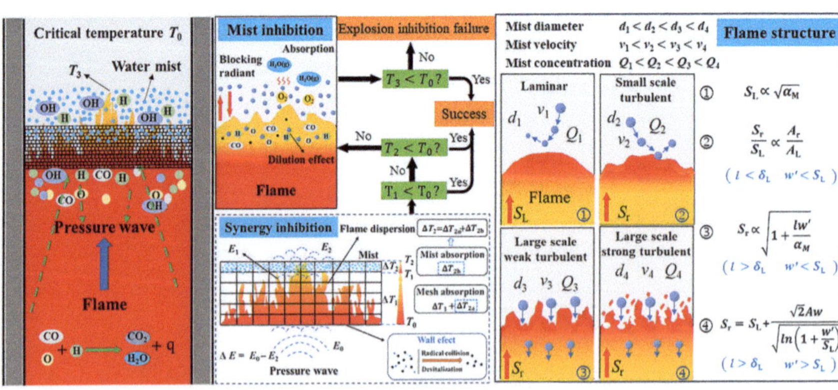

Fig. 6.14 Schematic diagram of synergistic inhibition mechanism

processes between the flame and wire mesh. Meanwhile, the presence of wire mesh would cause the obvious increases in the contact time and area between the flame and water mist. The mutual promotion of the two led to an increase in inhibition effect for explosion flame. As the ejected burned gas had an enough energy and the synergistic effect did not reduce the flame temperature below the critical value, the high temperature burned gas would be ejected from mesh layer. And the gas in the upper part would be ignited and the flame could continue to spread even if the space containing the water mist. Besides, it was also related to the mesh structure of water-containing layer and its thickness.

The flame would pass through the wire mesh and ignite the premixed gas of upper end after explosion-retardant failed. Afterwards, the flame exhibited different inhibitory effects due to the synergistic effect of mist. The flame entered the mist zone after passing through the wire mesh, and the temperature of flame front could be greatly reduced by heat absorption of liquid mist. Meanwhile, the water vapor would dilute the concentration of premixed gas at the reaction zone and reduce the ability of transfer heat through the thermal radiation. Besides, the highly active free radicals (H, O, and OH) during the syngas explosion would collide with the water molecules, reducing the number and energy of free radicals in the chain reaction process, thereby reducing the intensity of explosion reaction. However, the mist would also cause varying degrees of disturbance on the flame front and flow field, promoting the explosion reaction and leading to varying degrees of enhancement, and was related to the mist parameters (the mist diameter, concentration and its motion velocity). The diameter, concentration and its motion velocity of mist was altered with the changes of nozzle type and spray pressure, causing varying degrees of disturbance on the flame front. The flame front appeared four structures. Namely, the laminar flame, small-scale weakly turbulent flame ($l < \delta_L, w\prime < S_L$), large-scale weakly turbulent flame ($l > \delta_L, w\prime < S_L$), and large-scale strongly turbulent flame ($l > \delta_L, w\prime > S_L$). (Turbulent scale: l; Laminar flame thickness: δ_L; Fluctuation velocity: $w\prime$; Laminar flame rate: S_L). Under the laminar flame structure, the structure of flame front could not be altered after contacting mist. And the mist could absorb the heat released by the thermal radiation of flame front and disperse above the flame front after vaporization. As the diameter and velocity of mist increased, the momentum and pulsation intensity of mist acting on the flame front were increased. Under the action of mist, the flame front was no longer smooth and became a regular "wave" structure. At this moment, the mist and flame front could achieve good heat exchange, greatly reducing the explosion reaction rate. As the diameter and velocity of mist continued to increase, the mist with a certain pulsating intensity would penetrate the flame front and cause irregular deformation of its structure. The mist could not fully exert its heat absorption effect on the flame front and cause an increase in its surface area, further promoting an increase in the heat exchange rate between burned and unburned gases. As the flame was subjected to the mist of high pulsating intensity, the turbulent vortices would break through the flame front and even lead to the discrete structures. Burning and pulsating fresh air masses could quickly enter each other. According to the Darantov turbulent flame propagation formula, the flame propagation velocity could be improved sharply with the increase of pulsation velocity. Meanwhile, the effect of mist on the flame front was related to the mist parameters and was affected by three comprehensive actions.

6.3 Conclusions

The experiment investigated the synergistic effect of resistance/inhibition on the flame propagation characteristics of syngas explosions. The correspondence and difference between the explosion parameters at the upper and lower ends of wire mesh during success and failure were analyzed. The influence regular and mechanism of nozzle type and spray pressure on explosion inhibition parameters were clarified. The conclusion are as follows:

(1) Under the synergistic effect condition, a clear corresponding relationship appeared between flame propagation and pressure rise. As the explosion-retardant failed, the flame underwent three accelerated propagation processes. Namely, the self-acceleration, wire mesh acceleration and turbulent acceleration. The reflected pressure wave generated at the vessel upper end pass through the wire mesh, resulting in a sharp decrease in the rate of pressure rise at the upper end and even a larger negative value. The pressure rise at both ends was slightly earlier compared with the failure, and the flame propagation appeared two acceleration processes and the velocity history showed a double peak value. The pressures at both ends only presented an acceleration process, and the peak value was evidently decreased. After the inhibition was successful, the explosion intensity was also obviously decreased.

(2) The changes of nozzle type and spray pressure resulted in an obvious difference in the mist parameters, which affected the synergistic result. Synergies effect could evidently reduce the intensity of syngas explosion. As nozzle type increased, the synergetic effect was continuously enhanced. The velocity history and explosion intensity showed a significant decrease trend as the mist concentration increased. However, the synergetic inhibition appeared a change trend of failure–success–failure as the spray pressure increased. The corresponding explosion parameters also presented a trend of decreasing first and then increasing.

(3) After adding mist, the mist could exhibit a significant inhibition effect. However, the flame front appeared four structures affected by water mist. Namely, the laminar flame, the small-scale weakly turbulent flame, the large-scale weakly turbulent flame, and the large-scale strongly turbulent flame. The intensity of mist action on flame was related to its turbulent scale and pulsation velocity and was affected by mist parameter, further affecting the inhibition effect.

For better achieve explosion protection during the production and utilization of energy, three suggestions were proposed base on this experiment: (1) The comprehensive effect of changes in mist parameters on flame should be considered to achieve more effective explosion inhibition. (2) Inhibitors that could improve the physical and chemical effects of mist should be dissolved in water to improve the inhibition effect of spray. (3) To better leverage synergies, the structural parameters and materials of explosion-resistant system should be optimized.

References

1. Cao XY, Zhou JY, Zhou X, Wang Z, Wang ZR, Sheng YX. Experimental research on the synergy effect of resistance/inhibition on the syngas explosion. Fuel. 2024;363:130995.
2. Fuss SP, Chen EF, Yang W, Kee RJ, Williams BA, Fleming JW. Inhibition of premixed methane/air flames by water mist. Proc Combust Inst. 2002;29(1):361–8.
3. Zhang B, Yang ZZ, Leo YD. On the dynamics of drifting flame front in a confined chamber with airflow disorder in a methane-air mixture. Fuel. 2024;357:129761.
4. Li WY, Zou C, Yao H, Lin QJ, Fu R, Luo JH. An optimized kinetic model for H_2/CO combustion in CO_2 diluent at elevated pressures. Combust Flame. 2022;241:112093.

Chapter 7
Research on the Synergy Effect of Water Mist Containing Additives and Flame-Retardant System on the Syngas Explosion Inhibition

7.1 Experimental Apparatus and Procedures

The explosion resistance/inhibition apparatus is presented in Fig. 7.1. The detailed instruction had been stated in the previous experiment work. Here, only the explosion-inhibition system (Pressure spray system) and flame-retardant system (Metal wire mesh system) were introduced. The pressure spray system included a stainless steel water storage container, a solenoid valve, a check valve and a pressure nozzle. The deionized water was loaded into a storage tank and the additive was dissolved into it. The inhibition condition was adjusted by changing the additive condition (its type and mass fraction) and the spray pressure. By controlling the solenoid valve, the spray moment and duration were adjusted, thereby changing the mist concentration at the upper part of flame-retardant system. The flame-retardant system with square cross-section (100 mm × 100 mm) was mainly composed of two layer pressure plates with 3 mm thickness, multi layer wire mesh and four supporting rods. The wire mesh was compressed inside two pressure plates by adjusting nut height. The flame-retardant system was installed on the supporting rod with a determined position (H = 650 mm). To guarantee the fixed compression strength of wire mesh under the same working condition, four equally high metal gaskets were installed between the two pressure plates.

Besides, the high-frequency thermocouples with 2000 °C range and the pressure transmitters with 50 kHz response frequency (Range: 0–2.5 MPa; Accuracy: 0.5% FS) were respectively installed on wall surfaces on both sides of the vessel at a distance of 50 mm from the lower and upper interfaces of the flame-retardant system to gather local temperature and pressure data under explosion resistance/inhibition success and failure conditions. The ignition method and the high-speed camera [(f) = 4000 fps] were introduced in the previous work. The position of flame front was determined based on pixel coordinates of post-processing system and the flame propagation velocity was calculated by proportional conversion. The sequence and

Z. Wang and X. Cao, *Flame Resistance Characteristics and Its Influence Mechanism*,
https://doi.org/10.1007/978-981-95-6119-3_7

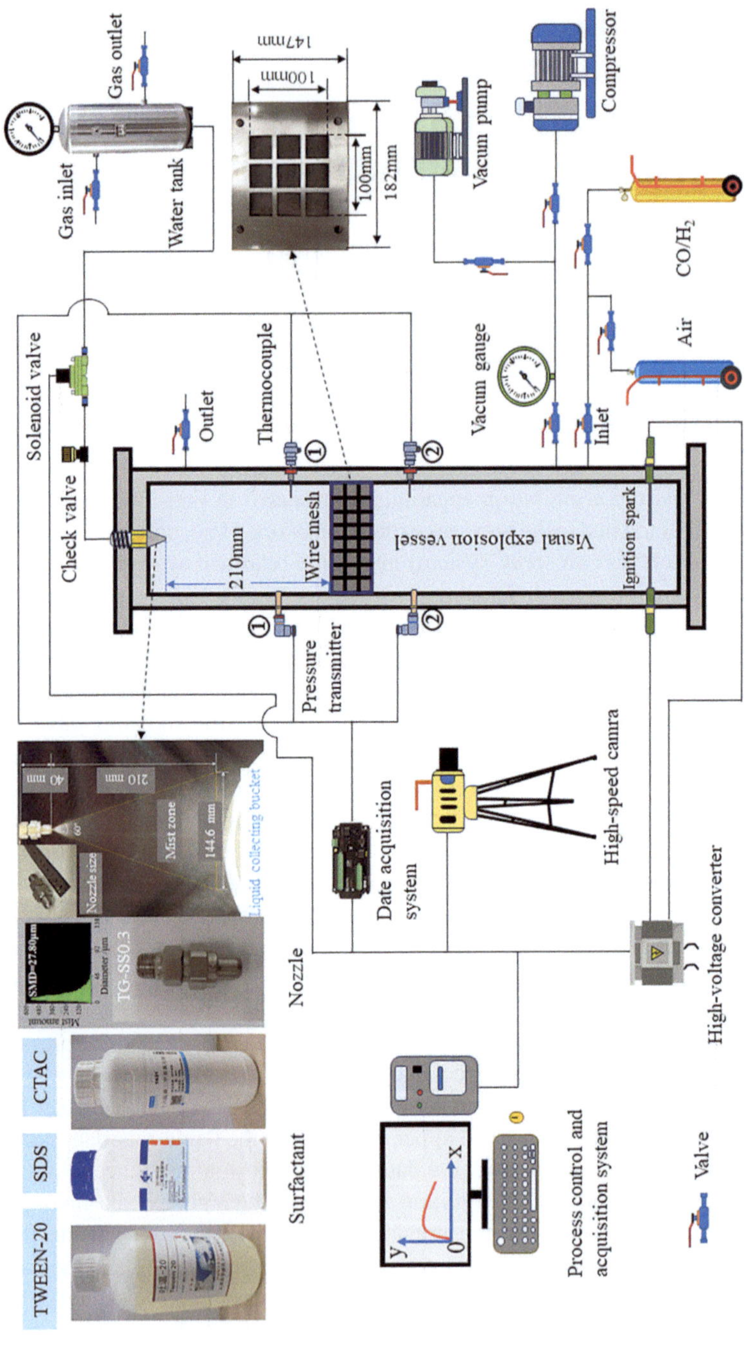

Fig. 7.1 Schematic diagram of experimental setup

Table 7.1 Detailed information of different additives

Additive	Structural formula	CAS	Type	Manufacturer
Tween-20	$C_{26}H_{50}O_{10}$	9005-64-5	Nonionic	Tianjin Hua Sheng Chemical Reagent Limited Company
SDS	$C_{12}H_{25}NaO_4S$	151-21-3	Anionic	
CTAC	$C_{19}H_{42}ClN$	112-02-7	Cationic	

time of spray, ignition and data storage were realized by the program control system (200 kHz/per channel) [1].

A solid conical nozzle [*Spraying System Company*; (TG-SS 0.3)] was selected to produce the mist and the mist diameter at a distance of 30 mm below the nozzle under 0.60 spray pressure was measured through the phase doppler particle analyzer [Sauter Mean Diameter (SMD) = 27.80 μm]. It indicates the diameter of a sphere with the same surface area as the actual liquid drop. The motion velocity of mist was 9.20 m/s and its spray amount was 3.21 g/s. In the experiments, three additives [$C_{26}H_{50}O_{10}$ (Tween-20), $C_{12}H_{25}NaO_4S$ (SDS), and $C_{19}H_{42}ClN$ (CTAC)] were added to deionized water to form the solutions of five mass fractions ($\omega = 0.1\%$, 0.2%, 0.3%, 0.4%, and 0.5%) for synergy inhibition, as can be seen in Table 7.1. The premixed gas (H_2:CO = 1:1) with a concentration of 25% volume fraction was prepared inside the vessel. First of all, the flame-retardant system with determined the structure parameters [Layer number (n) = 25; Mesh number (m) = 40] was installed inside the vessel. To ensure the validity of data, three repeated experiments were conducted under the same operating conditions. By calculation, the deviations of pressure peak (ΔP_{max}), velocity peak (v_{max}) and temperature peak (T_{max}) were 5.8%, 3.9% and 8.6%, respectively.

7.2 Results and Discussions

7.2.1 Analysis of Synergy Effects

Figure 7.2 shows the correspondence between flame propagation and explosion parameter as the synergy was failed (TWEEN-20 additive; $t_{sp} = 400$ ms; $\omega = 0.2\%$). It was apparent from the figure that the high temperature flame could pass through the flame-retardant system and the mixed gas of upper part was rapidly ignited. In the early stages, the flame propagated in a "spherical" structure and the pressure wave generated at its front propagated upwards along the axis of vessel after ignition. The flame exhibited the weak propagation ability in the early stages, resulting in that the pressure did not show an evident rise process. Constrained by the wall surface, the flame changed from the "spherical" to "finger" structures and its propagation velocity was dramatically improved. The lower pressure history started to rise at 7.00 ms. With the acceleration of flame propagation, the increase in pressure became significant. The velocity history emerged the I-th peak value at 13.75 ms [$v_{(1)}$ =

43.2 m/s]. In comparison, the moment of upper pressure rise was slightly delayed ($t = 9.50$ ms) and its rise rate was obvious smaller than that of lower end. Its velocity was evidently decreased as the flame curvature area was continuously enlarged. At 17.00 ms, the flame approached the flame-retardant system and appeared the axial stretching phenomenon in the central area. At this point, the velocity history emerged the II-th peak value [$v_{(2)} = 41.7$ m/s]. Thereafter, the flame entered the flame-retardant system at 17.75 ms and stayed for 7.00 ms before passing through its upper interface and igniting the upper mixed gas at 24.75 ms. At 18.75 ms, (dP/dt) history of lower end appeared the I-th peak value. Nevertheless, it was slightly greater than that of upper end and the appearance moment of I-th peak value at the upper part was also delayed ($t = 19.50$ ms). After passing through the upper interface, the discrete flame exhibited the acceleration phenomenon and the velocity peak value appeared at 27.25 ms ($v_{max} = 131.6$ m/s). And the peak value of (dP/dt) history appeared at 29.00 ms. Thereafter, ΔP_{max} appeared at 32.00 ms. Blocked by the vessel top end, the reflected pressure wave passed through the flame-retardant system, further resulting in a second rise in the lower pressure history and presenting the II-th peak value at 33.25 ms. Afterwards, the upper pressure was sharply decreased, and its (dP/dt) history even appeared a remarkable negative value [$(dP/dt)_{min} = -28.9$ MPa s^{-1}]. As the flame extinguished, the rise rate of pressure was greatly reduced. At 39.50 ms, the lower pressure reached the maximum value and it was markedly reduced than the upper end. Subsequently, the spreading flame was quickly extinguished and the pressure was also rapidly reduced.

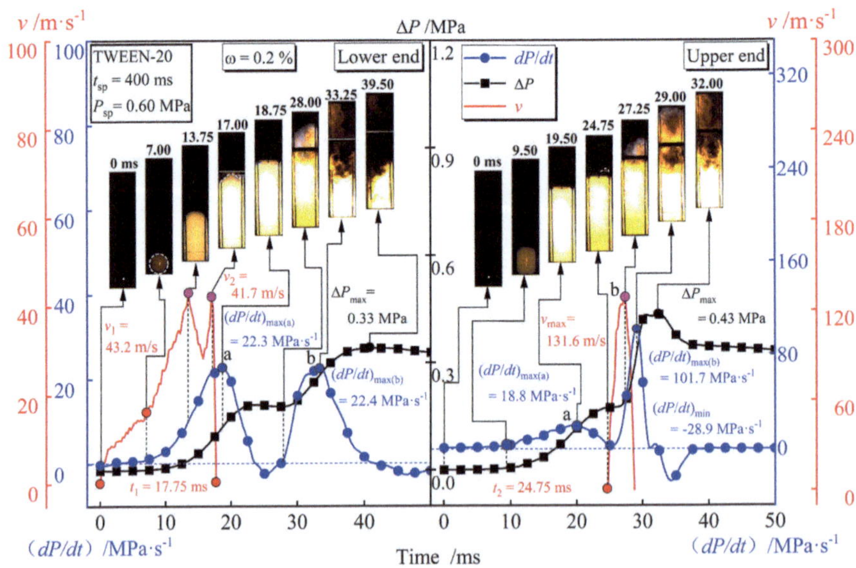

Fig. 7.2 Correspondence between flame propagation and explosion parameter under the synergy failure condition (TWEEN-20 additive; $t_{sp} = 400$ ms; $\omega = 0.2\%$)

Figure 7.3 presents the correspondence between explosion resistance/inhibition parameters under the synergy success condition (CTAC additive; t_{sp} = 400 ms; ω = 0.2%). Here, the flame did not pass through the flame-retardant system to ignite the flammable gas of upper end. But an apparent correspondence was still presented between the two. The lower end flame underwent two identical acceleration processes, and there was no obvious difference in the peak value of velocity history compared with the failure [$v_{(1)}$ = 43.4 m/s; $v_{(2)}$ = 41.9 m/s]. But its appearance moment was slightly ahead of schedule. Compared with the failure, the moment for pressure rise was evidently delayed as the flame retardant was successful (t = 10.25 and 12.50 ms). And the lower pressure only presented a rise process and ΔP_{max} was 0.19 MPa, which was far less than that of flame retardant failure. For upper end, the drop extent of ΔP_{max} was more outstanding (from 0.43 to 0.17 MPa). Additionally, the moment for lower end ΔP_{max} appearance was advanced owing to the enhancement of flame-retardant system reflection action (t = 29.50 ms). But there was no significant difference in the appearance moment of ΔP_{max} at the upper end. Correspondingly, the pressures of two ends all presented an acceleration upward process. And the peak value of (dP/dt) history at upper end were decreased from 101.7 to 17.9 MPa/s. It fully demonstrates that the synergy effect of metal wire mesh and mist containing CTAC additive could prominently weaken the intensity of syngas explosion and the pressure drop of upper end performed more remarkable.

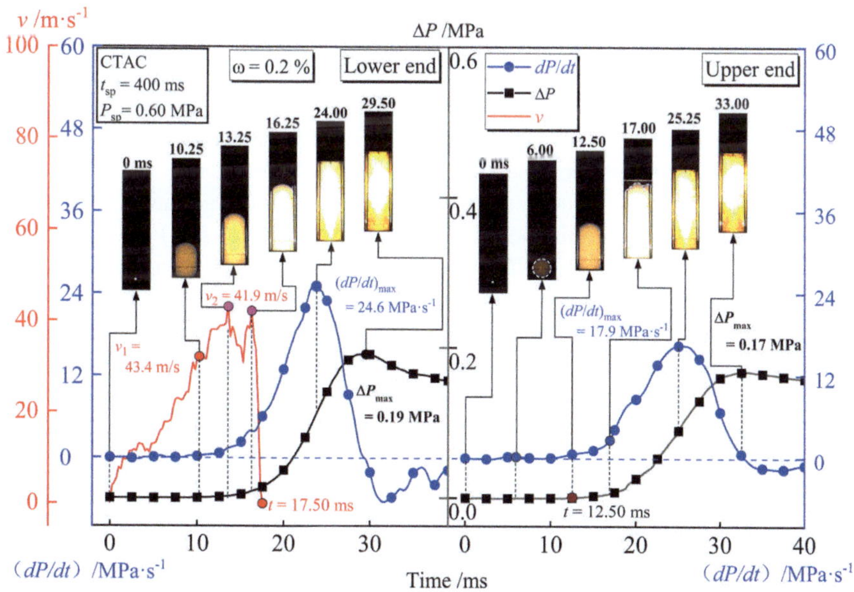

Fig. 7.3 Correspondence between flame propagation and explosion parameter under the synergy success condition (CTAC additive; t_{sp} = 400 ms; ω = 0.2%)

7.2.2 Analysis of Flame Propagation Characteristics

Figure 7.4 presents the influence of additive type on the flame propagation (t_{sp} = 400 ms; ω = 0.2%). And the corresponding velocity history was showed in Fig. 7.5. It clear that the order of three additives in weakening flame propagation was CTAC > SDS > TWEEN-20. The additive type presented a more significant impact on the upper end flame than lower end. In the early stages of development, the smooth flame all accelerated propagation and transformed from the "spherical" to "finger" structure. Meanwhile, the lower end flame all presented two acceleration process under different additives, and there was no present an obvious difference. As it approached the flame-retardant system, the flame front appeared the axial stretching phenomenon and exhibited a second acceleration process [2]. Under pure water mist condition, the gas of upper end was firstly ignited after staying for 5.50 ms inside the flame-retardant system. Under the actions of mist containing TWEEN-20 and SDS additives, the flame passed through flame-retardant system and ignited the above premixed gas successively. The stagnation time was prolonged to 7.00 ms and 9.00 ms respectively. The flame exhibited the obvious turbulence and underwent the third acceleration process. Besides, the flame failed to penetrate the flame-retardant system to ignite the above flammable gas, indicating the synergy success under the action of mist containing CTAC additive. It also fully demonstrates that the weakening effect of mist on the flame would be obviously improved after adding additive and the enhancement effect of CTAC additive was most prominent. Simultaneously, the velocity history of upper end occurred obvious change, and the peak values were decreased from 144.1 m/s (Pure mist) to 131.6 m/s (TWEEN-20) and 119.4 m/s (SDS), respectively. Besides, the weakening effect of the three additives on the flame propagation at the upper end was enhanced and all showed the same intensity order as the spray amount increased.

Figure 7.6 illustrates the influence of CTAC mass fraction on flame propagation process (t_{sp} = 200 ms). Combined with the velocity history (see Fig. 7.7), it can be inferred that the weakening effect of mist containing additive on the flame was mainly reflected in the velocity history of wire mesh upper end, while the lower end flame did not show the significant change with time. When the flame retardant failed, the flame emerged the three similar acceleration processes. In the early stages, the flame experienced its first acceleration. At 19.00 ms, the flame stretched and underwent the second acceleration owing to the reflex action of flame-retardant system. Subsequently, the smooth flame was discretized by wire mesh and exhibited the third acceleration under the failure condition. Compared with the pure water mist, the weakening effect of mist on the flame propagation would be improved after the adding CTAC. Within the range of $0.1\% \leq \omega \leq 0.5\%$ CTAC mass fraction, the stagnation time of flame inside the flame-retardant system showed a variation regular of first rising and then reducing as the CTAC mass fraction increased, and the upper end velocity history also presented the similar change regular. As ω = 0.2%, the inhibition effect reached the best. Especially, the height of flame front was the lowest by comparison and the flame stagnation time was the longest (Δt = 5.75 ms). Besides,

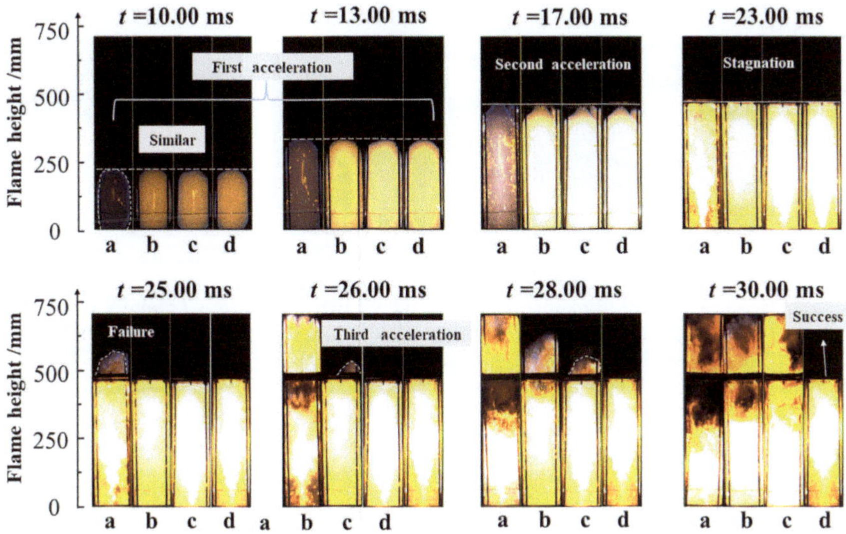

Fig. 7.4 Influence of additive type on the flame propagation process (t_{sp} = 400 ms; ω = 0.2%)

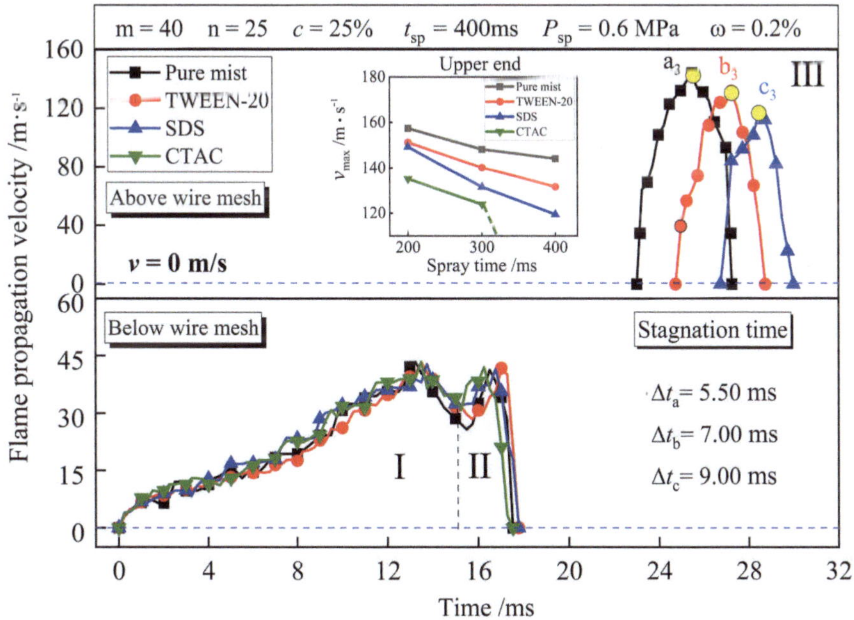

Fig. 7.5 Influence of additive type on the flame propagation velocity history

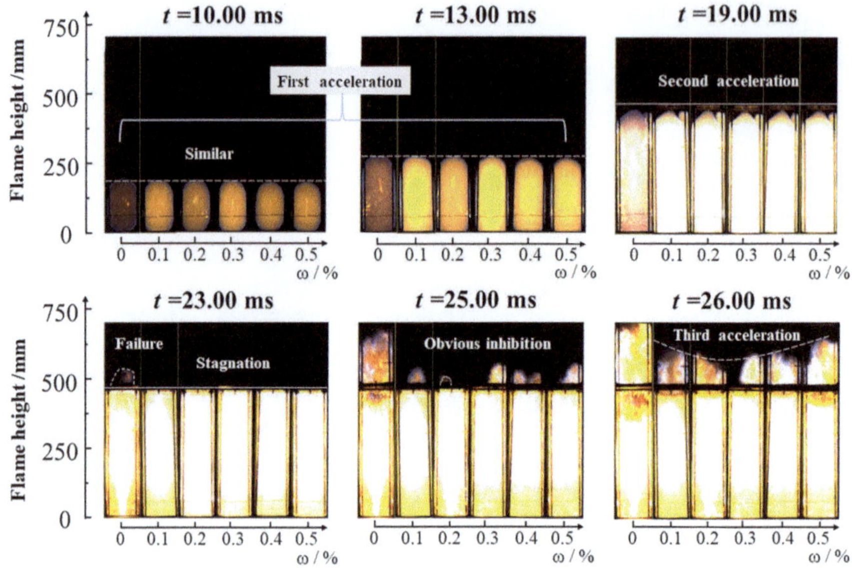

Fig. 7.6 Influence of CATC additive mass fraction on the flame propagation process ($t_{sp} = 200$ ms)

v_{max} of upper end was gradually decreased as the spray time increased based on the 3D surface graph. As the mass fraction increased, there was a change regular of first reducing and then rising. And the velocity history of upper end reached the minimum value under 0.2% mass fraction. It was because that the CTAC additive could reduce the surface tension of mist and exhibit good foaming performance, leading to the appearance of optimal inhibition effect at this additive mass fraction.

7.2.3 Analysis of Explosion Pressure

The above indicates that the type of additive could evidently affect the weakening effect of mist on the flame propagation, and this inhibition extent was also reflected in the pressure. Figure 7.8 illustrates the influence of additive type on the explosion pressure ($t_{sp} = 400$ ms; $\omega = 0.2\%$). As the flame retardant failed, the additive type did not present an obvious effect on the lower end pressure, while the upper end pressure could be greatly reduced. And the order of inhibition ability was CTAC > SDS > Tween-20. Compared with the absence of additives, ΔP_{max} of upper end under mist containing TWEEN-20 and SDS additives were reduced respectively by 8.5% and 17.02% and their appearance moments were also delayed. In particular, the flame was successfully inhibited under mist containing CTAC additive, and the decrease extent of ΔP_{max} reached 63.8% and 40.6% under both upper and lower ends.

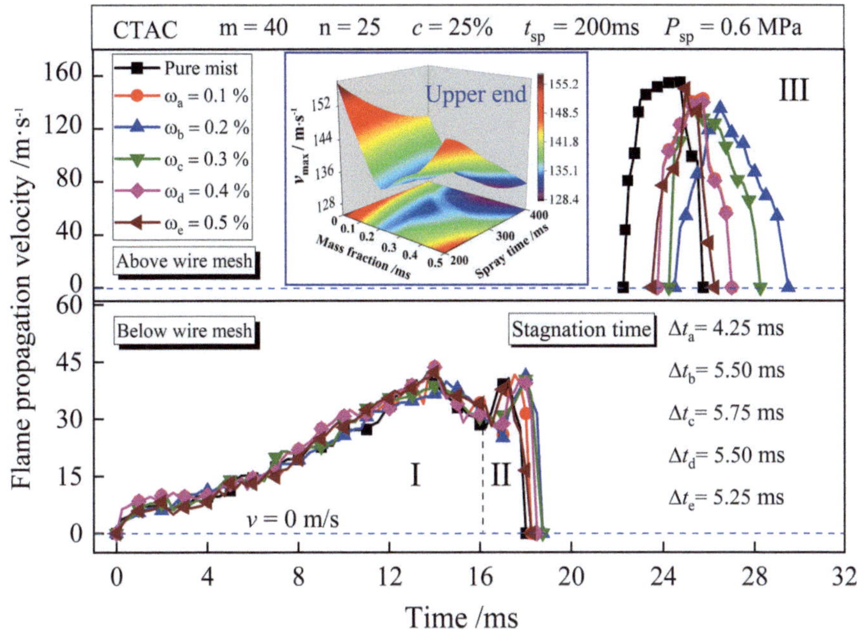

Fig. 7.7 Influence of additive mass fraction on the flame propagation velocity history

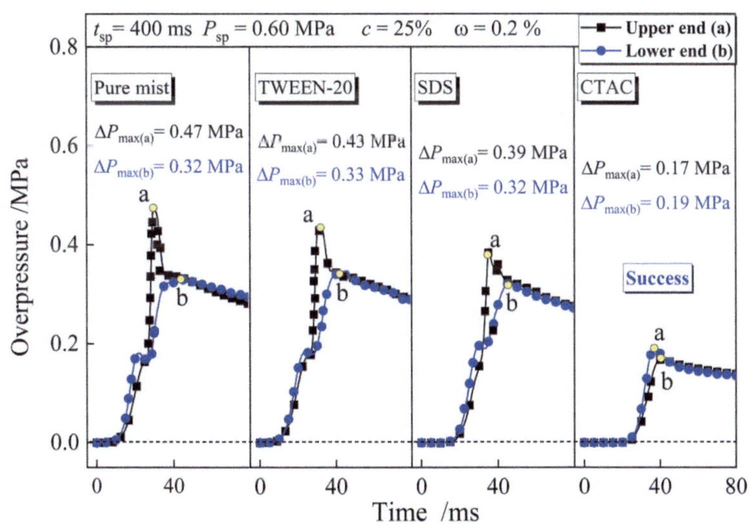

Fig. 7.8 Effect of additive type on the pressure history (t_{sp} = 400 ms; ω = 0.2%)

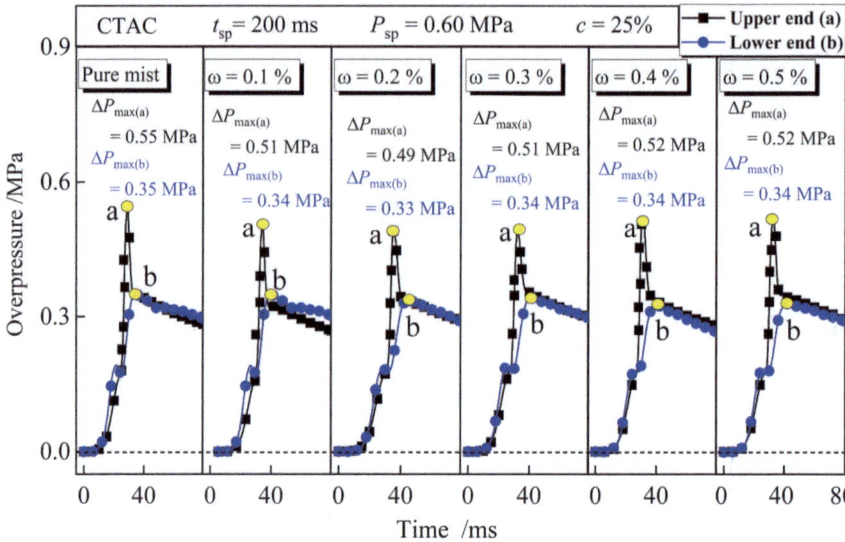

Fig. 7.9 Effect of additive mass fraction on the pressure history (t_{sp} = 200 ms; P_{sp} = 0.60 MPa)

Meanwhile, the mass fraction of additive would also obviously affect the pressure, as shown in Fig. 7.9. Especially for the impact on the upper end pressure. And it was higher than the lower end pressure. The weakening effect of mist on explosion could be enhanced after adding a small amount of CTAC additive. Within the range of 0.1% $\leq \omega \leq$ 0.5% CTAC mass fraction, ΔP_{max} of upper end displayed a variation regular of first decreasing and then slightly increasing. Under 0.2% CTAC mass fraction, the weakening effect was the strongest and ΔP_{max} was decreased by 10.9% compared with the pure water mist. Afterwards, ΔP_{max} of upper end tended to flatten but it was still decreased compared with the absence of CTAC additive as the solution mass fraction increased. It was because that the weakening effect of mist containing CTAC was prominently affected by the comprehensive effect of the solution surface tension and foam volume [3]. Besides, the pressure presented the same change regular as the spray time increased. Differently, the mass fraction of Tween-20 additive reaching the optimal inhibition effect was 0.3%. But the optimal inhibition effect was 0.4% for SDS.

7.2.4 Analysis of Pressure Rise Rate

Figure 7.10 displays the influence of additive type on (*dP*/*dt*) history (t_{sp} = 400 ms; ω = 0.2%). The (*dP/dt*) could directly reflect the weakening extent of additive on syngas explosion. It clears that the influence of additive type on the explosion reaction rate emerged the same variation regular as the pressure history, and the mist containing CTAC exhibited the strongest impact on the explosion reaction. Under the failure conditions, (dP/dt) history emerged two peaks, and (dP/dt) of upper end was larger compared with the lower end and its change was also more significant. The mist containing TWEEN-20 and SDS could evidently reduce the $(dP/dt)_{max(a)}$ of upper end, resulting in a decrease extents of 27.6% and 46.3%, respectively. Simultaneously, the II peak value of upper end was higher than I peak value, yet the peak value of lower end (dP/dt) history was similar. Because of the reflection action of vessel top end, the pressure wave propagated backwards and passed through the flame-retardant system, leading to a large negative value in (dP/dt) history of upper end, and the influence of additive type on it also existed significant difference. As the flame retardant was successful, only one peak appeared in the (dP/dt) history. After adding CTAC additive, the mist of solution could greatly reduce the explosion intensity. Under inhibition success condition, $(dP/dt)_{max}$ of upper end was decreased by 87.05% than without addition. And the same pattern of change was observed under different spray amounts.

Figure 7.11 gives out the influence of CTAC solution mass fraction on the (dP/dt) history ($t_{sp} = 200$ ms). As the solution mass fraction increased, the $(dP/dt)_{max}$ at both ends of flame-retardant system presented the change regular of first reducing and then rising. As $\omega = 0.2\%$, both them reached their minimum values and $(dP/dt)_{max}$ of upper and lower ends were reduced respectively by 20.2% and 47.0%. Accordingly, the negative value of (dP/dt) history at the upper end also showed the same trend of change. And its absolute value was the smallest under 0.2% mass fraction. It was

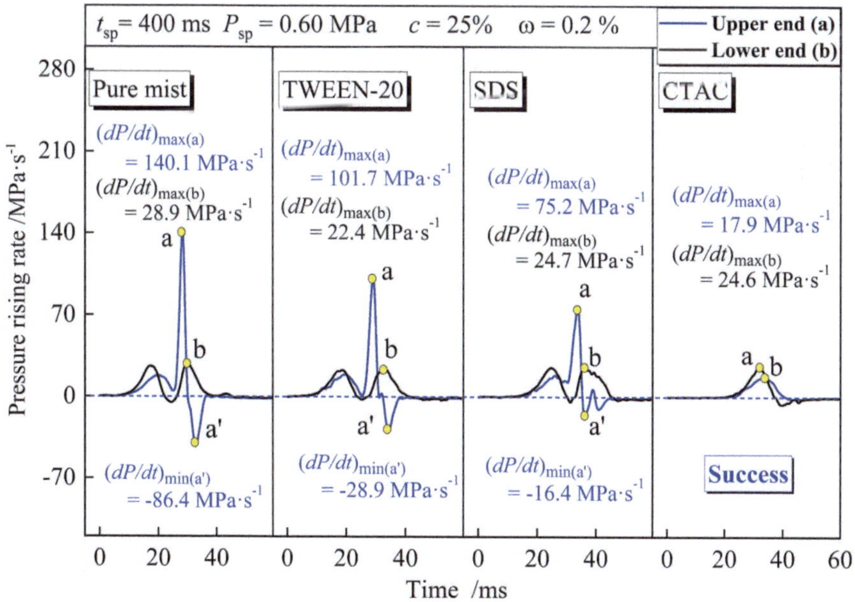

Fig. 7.10 Effect of additive type on (dP/dt) history ($t_{sp} = 400$ ms; $\omega = 0.2\%$)

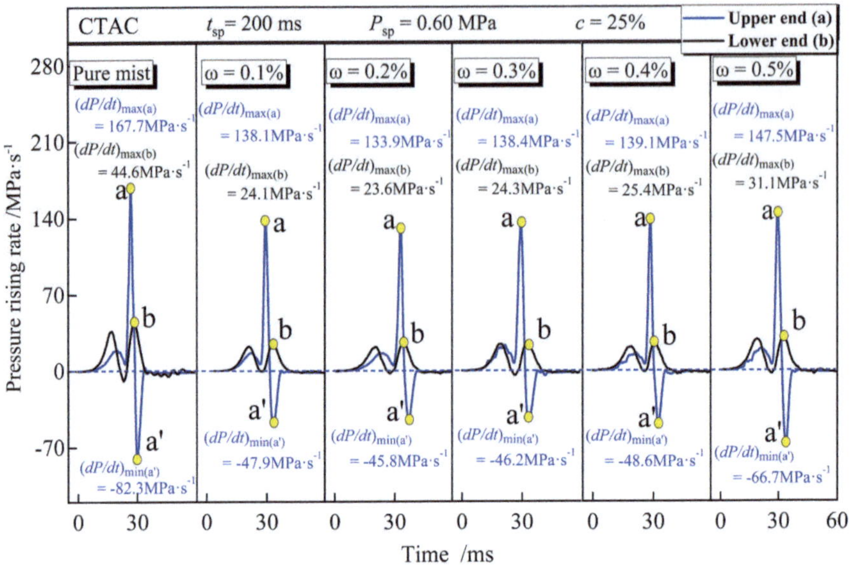

Fig. 7.11 Effect of additive mass fraction on (*dP/dt*) history ($t_{sp} = 200$ ms; $P_{sp} = 0.60$ MPa)

decreased by 44.3% compared with the absence of additive. Besides, the effects of mist containing Tween-20 and SDS additives on ΔP_{max} and $(dP/dt)_{max}$ was similar to that of CTAC additive with increasing mass fraction, and all displayed a change law of first reducing then tending to flatten out. Differently, the mass fraction of Tween-20 additive reaching the optimal inhibition effect was 0.3%. But the most significant inhibition effect was 0.4% for SDS.

7.2.5 Analysis of Flame Temperature

Figure 7.12 gives out the effects of additive type and mass fraction on flame temperature. The upper end temperature was higher compare with the lower end under the failure condition, and T_{max} at both ends of flame-retardant system could be reduced after adding additive. Compared with pure water mist, T_{max} of upper end was decreased by 13.4% after adding SDS additive, and its decrease extent was evident higher than that of TWEEN-20 additive. Nevertheless, the decrease extent in the temperature of lower end under two additive conditions did not present the obvious different, but effect of SDS additive was still slightly lower than TWEEN-20 additive.

After adding CTAC additive, the synergy was extremely successful. Under this condition, the drop extent of upper end T_{max} was more remarkable and was smaller than that of lower end (from 1649 to 171 °C), causing the flame to fail to ignite the above flammable gas. It can be inferred that the type of additive could affect the

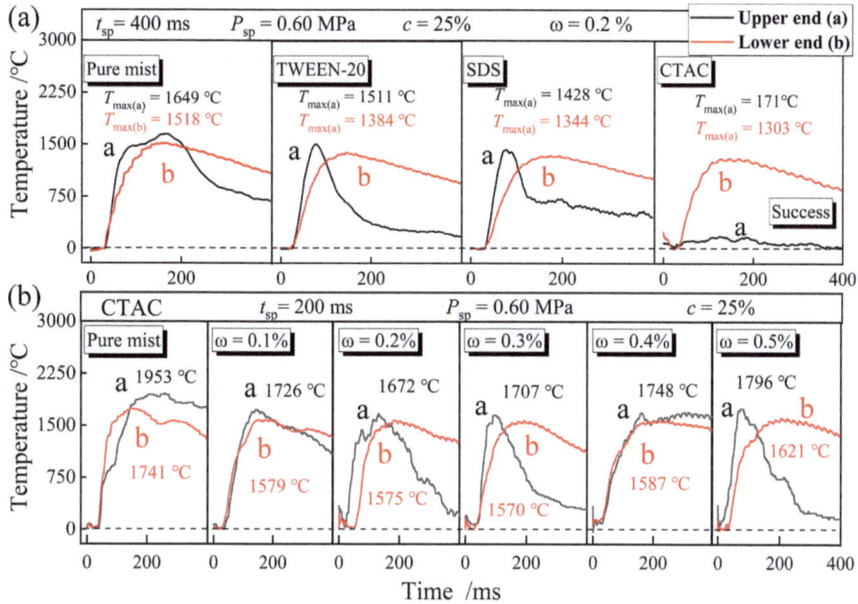

Fig. 7.12 Effect of additive on flame temperature history

endothermic effect of mist on flame and their order was CTAC > SDS > Tween-20 > pure water mist. Concurrently, the temperature was greatly decreased after adding CTAC additive. Within the range of $0.1\% \leq \omega \leq 0.5\%$ CTAC mass fraction, T_{max} displayed a change law of first reducing and then rising as the CTAC mass fraction increased. Under different additives, the flame temperature all exhibited the same regular as the pressure changed. However, T_{max} at both ends were the lowest under 0.2% CTAC mass fraction (T_{max} = 1672 and 1575 °C).

7.2.6 Analysis of Synergy Mechanism

Figure 7.13 presents the synergy of flame-retardant system and mist containing additives for syngas explosion. After ignition, the flame propagated upward and accompanied by a large amount of active free radical chain reaction processes ($O + H_2O = OH + OH$, $CO + OH = CO_2 + H$, $CO + O_2 = CO_2 + O$, $H + O_2 = O + OH$, $2OH = O + H_2O$, $H + O_2 = OH + O$, $OH + H_2 = H + H_2O$). The part of pressure wave generated was reflected due to the obstruction of flame-retardant system, and part of pressure wave energy would be absorbed as it come into contacted with the flame-retardant unit, further greatly weakening the explosion energy. Simultaneously, the continuous and smooth flame was discretized into a large number of discontinuous tiny flames by internal narrow channel, further increasing the heat transfer area and rate. The destruction of high-energy free radicals by wire mesh caused to the significant reduce in the number and collision probability of active free radicals in the narrow channel, further greatly weakening the intensity of chain reaction and thereby reducing the ability of flame propagation. In addition, a layer of mist was attached at the upper interface of flame-retardant system. As it passed through upper interface, the mist could also reduce the flame temperature by absorption heat effect. The vapor dispersed inside the flame-retardant system, further diluting and blocking the spread of flame. And the different additives would exhibit varying degrees of weakening. Under the above comprehensive effects, the flame temperature was decreased from T_0 (wire mesh lower interface) to T_2 (wire mesh upper interface). If T_2 was less than the critical flame temperature T', its energy was not sufficient to ignite the syngas at the upper end. Namely, flame-retardant success. Otherwise, the flame would enter the mist zone at the upper end.

As the flame retardant failed, the flame passed through the flame-retardant unit and could ignite the premixed gas of upper end. Here, the temperature of flame front could be decreased through endothermic effect, and the concentration of premixed gas would be diluted by the vapor generated from high-temperature gasification, further reducing the ability of transfer heat to unburned gas by the thermal radiation. Besides, the additive could affect the degree of weakening by altering the physical properties of mist. After adding additive, the surface tension of mist was reduced to varying degrees (SDS < Tween-20 < CTAC < Pure mist) [4]. It was due to existence of hydrophobic and hydrophilic groups in additives. After being dissolved in water, the hydrophobic groups aggregated towards the gas phase side, while the hydrophilic groups aggregated towards the liquid phase side. It caused the intermolecular force acting on the liquid phase in the vertical direction to cancel out each other and led to a decrease in surface tension of the liquid [5]. The reduce of surface tension would decrease the diameter of mist under the same spray condition, thus increasing the contact area between mist and flame and enhancing its heat absorption rate. In addition, the hydrophilic groups and water molecules attracted each other to form tiny "micelles". As the pressure spray was applied, it will absorb air and form bubble. After contacting with flame, it will further enhance the endothermic effect of mist on flame front and increase the dilution effect of premixed gas for the reaction zone.

Besides, the additives would also play an obvious chemical effect during the inhibition. Among the three additives, the non-ionic surfactant Tween-20 formed the ether bonds with H_2O molecules after it dissolved in water. Under high-temperature and pressure environments, the movement of Tween-20 molecules connected to water molecules by unstable hydrogen bonds intensified, causing the ether groups bound to break and detach from the water molecules. The broken ether group could capture H radicals of the chain reaction, leading to an increase in the consumption of active groups and a decrease in the rate of chain reaction during the explosion. The hydrolysis of anionic surfactant SDS could produce the Na^+ and $CH_3(CH_2)_{11}OSO_4{}^{-1}$. Among them, Na^+ participated in the radical reactions ($Na + OH + X = NaOH + X$, $NaO + H = Na + OH$, $NaOH + OH = NaO + H_2O$, $NaO + O = Na + O_2$, $NaOH + H = Na + H_2O$), consuming a large amount of H and OH radicals and to some extent interrupting the chain reaction process of syngas explosion. In addition, the

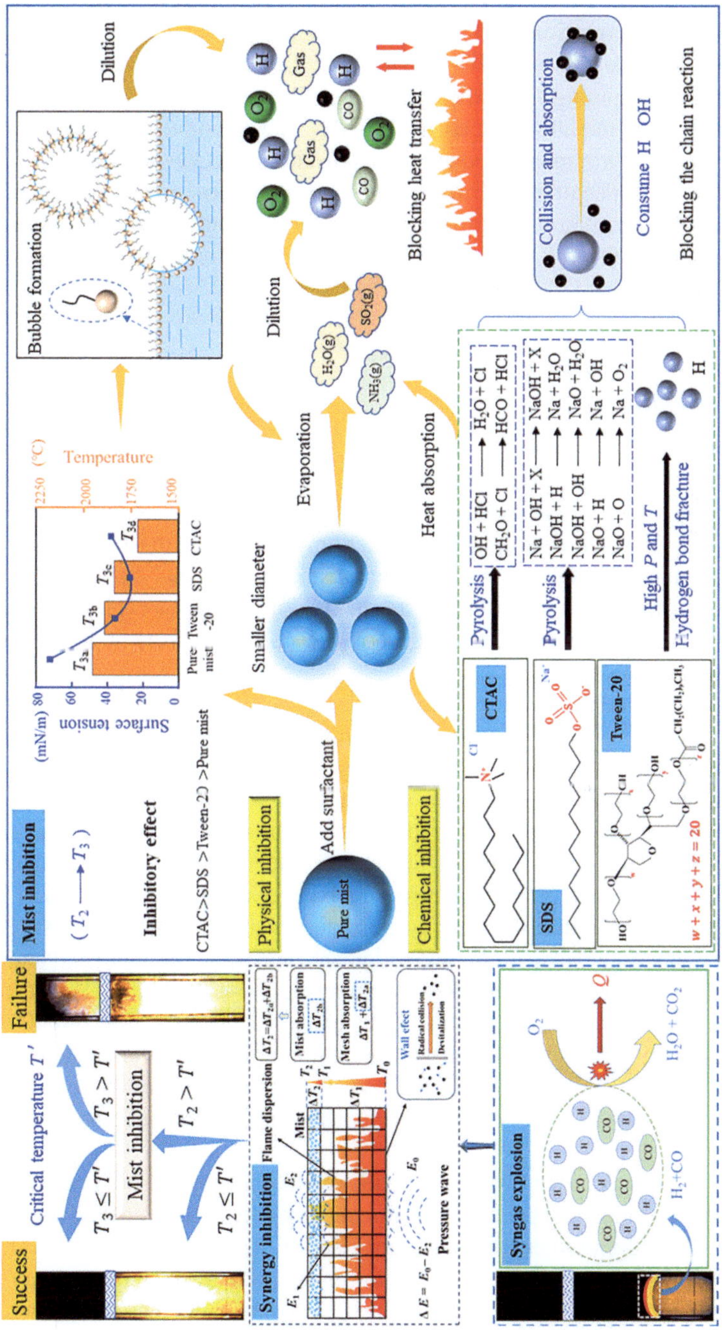

Fig. 7.13 Synergy mechanism diagram of explosion resistance/inhibition

$CH_3(CH_2)_{11}OSO_4^{-1}$ would produce gases such as H_2O and SO_2 under high temperature condition, thereby diluting the concentration of premixed gases on the flame front and reducing its reaction rate [5]. Similarly, the hydrolysis of cationic surfactant CTAC would produce Cl^- and $CH_3(CH_2)_{18}NH_3^+$. Cl^- could reduce the explosion reaction rate by consuming OH radical ($CH_2O + Cl = HCO + HCl$, $OH + HCl = H_2O + Cl$) [6]. And the alkyl in $CH_3(CH_2)_{18}NH_3^+$ was prone to decomposition at high-temperature environment, producing a small amount of CO_2 and vapor to dilute the flammable gas concentration. Under the combined physical and chemical effects mentioned above, the mist containing additive could evidently reduce the intensity of syngas explosion. Overall, the order of inhibition ability for mist containing three additives was CTAC > SDS > Tween-20 > Pure mist.

7.3 Conclusions

This experiment researched on the synergy of mist containing additive and flame-retardant system for syngas explosion inhibition inside the visualization closed vessel. By changing the type and mass fraction of additive, the changes in the explosion resistance/inhibition characteristic parameter were analyzed under opposite results. And the influence regular and mechanism of synergy on explosion parameter were revealed and the effective suggestion was proposed based on the research for practical application.

First, it all existed an evident correspondence between explosion resistance/inhibition parameter under the success and failure conditions. Compared with the synergy failure, the pressures of wire mesh two ends were markedly decreased and the moment of pressure rise was delayed as the inhibition was successful. The synergy effect could be enhanced after adding additive, and the intensity order of the three additives in weakening flame propagation was CTAC > SDS > TWEEN-20. By comparison, the influence of additive type on the upper end flame propagation was more significant. Meanwhile, the synergy presented a variation regular of first rising and then reducing as the additive mass fraction enhanced. v_{max} of upper end and the stagnation time of flame also exhibited a similar regular of variation. The stagnation time was prolonged from 5.50 to 9.00 ms under 0.2 mass fraction and 400 ms spraying time. For CTAC additive, the stagnation time was first prolonged from 4.25 to 5.75 ms and then reduced to 5.25 ms with increasing mass fraction. Besides, the synergy of the two on flame was continuously improved as the spray amount increased.

Second, the synergy of the two could greatly reduce the pressure of syngas explosion and the drop extent of upper end was more remarkable. Compared with the flame resistance failure, the moments for pressure rises at the two ends were evidently delayed as the synergy was successful. The lower end pressure only appeared a one-time increase process and was far smaller than the pressure under the synergy failure. The mist containing additive could greatly weaken the explosion intensity, and the influence of additive type on the pressure and its rising rate was similar with that of flame propagation change. However, the corresponding optimal inhibition concentration existed the obvious difference. Within the range of $0.1\% \leq \omega \leq 0.5\%$ mass fraction, the optimal inhibition concentrations for CTAC, Tween-20 and SDS additive mist solutions were 0.2%, 0.3%, and 0.4% mass fractions, respectively. And the pressure and its rising rate were continuously reduced as the spray amount enhanced.

Third, the mist containing additive could greatly reduce the flame temperature. Especially for the upper end flame. This demonstrates that the additive could improve the heat absorption effect of mist on the flame. The influences of additive type and mass fraction on flame temperature were the similar as the pressure history change. For 0.2% CTAC mass fraction, $T_{\max(a)}$ and $T_{\max(b)}$ were reduced respectively by 14.4% and 9.5% compared to pure water under 25% syngas concentration and 200 ms spray time. Especially, $T_{\max(a)}$ and $T_{\max(b)}$ were reduced respectively by 89.6% and 14.2% for 400 ms spray time. The synergy of mist and flame-retardant system was mainly attributed to the combination of chemical and physical effects, and the synergy could be enhanced after adding additive. And the addition of additive could also achieve the complete inhibition of syngas explosion.

To achieve effective protection against syngas explosion during industrial production and transportation, some suggestions are proposed according to the above conclusions: (1) Efficient and clean surfactants should be developed and optimal inhibition concentration should be determined to achieve effective inhibition for high activity gas explosion. (2) The equipment with large spray amount should be developed to realize fast and efficient spray. (3) The optimal prevention conditions for the synergy of explosion-resistant and explosion-inhibition two methods should be determined and clarified.

References

1. Cao XY, Zhou JY, Lu YW, Wang ZR, Wang Z. Research on the synergy effect of water mist containing additives and flame-retardant system on the syngas explosion inhibition. Appl Therm Eng. 2024;255: 124004.
2. Li WY, Zou C, Yao H, Lin QJ, Fu R, Luo JH. An optimized kinetic model for H_2/CO combustion in CO_2 diluent at elevated pressures. Combust Flame. 2022;241: 112093.
3. Xia X, Zhang HX, Xu GY, Tan YB, Zhang J, Lv X. Influence of CTAB and SDS on the properties of oil-in-water nano-emulsion with paraffin and span 20/Tween 20. Colloid Surf A Physicochem Eng Asp. 2013;418:60–7.
4. Jian HL, Li XR, Li DH. Study on fire-fighting effect and weakening of flame intensification of non-ionic liquid water mist. J Henan Polytech Univ. 2018;37:8–15.
5. Parra T, Castro F, Méndez C, Villafruela JM, Rodríguez MA. Extinction of premixed methane-air flames by water mist. Fire Saf J. 2004;39:581–600.
6. Hu ZW, Pei B, Xu MJ, Han YL, Lv H, Wu ZQ, Chen LW. Study on the inhibition effect and mechanism of N twin-fluid water mist with modified chloride compounds on LPG explosion. Energy J. 2024;291: 130394.

Part II
Corrugated Flame Arrester

Chapter 8 Research on the Resistance Characteristics of Corrugated Flame Arrester for Hydrogen Explosion

8.1 Experimental Apparatus and Procedures

The visualization experimental apparatus used to study the flame resistance characteristics of hydrogen explosions with a corrugated flame arrester is illustrated in Fig. 8.1. The main framework of this apparatus has been detailed in previous work [1]. This section focuses specifically on the corrugated flame-retardant system, which comprises the flame-retardant unit and its shell, upper and lower pressure plates, and four threaded support rods. The multilayer flame-retardant units are tightly stacked using the upper and lower pressure plates and are housed within the flame-resistant unit shell. The system is secured by the threaded support rods, with height adjustments made via nuts. By varying the ripple height and the number of flame-retardant units, the structural parameters can be altered. A self-processed flame-retardant unit with a thickness of 20 mm and the diameter of 110 mm was adopted.

The experimental conditions and parameters of the flame arrester are presented in Table 8.1. In order to prevent flame passing through the gap between the flame-retardant system and the inner wall surface, the gap was sealed by the high temperature explosion-proof sealing mud. The premixed gas inside the vessel was prepared according to Dalton's law of partial pressures. During the experiment, a high-speed camera ($f = 4000$ fps) recorded the transient transformation processes of flames at both ends of the flame arrester. Two high-frequency pressure transmitters [Range (A) = 0–2.5 MPa; Response frequency (f) = 200 kHz] and two S-type platinum–rhodium thermocouples were employed to collect pressure and temperature data at both ends of the flame arrester. The micro-thermocouple with S-type was welded from Pt and Pt/Rh-10% with a diameter of 50 μm and its measurement range in a short period of time reached 2000 °C. Additionally, ignition and data acquisition were controlled by a program and control acquisition system ($f = 200$ kHz/channel) developed using a custom programming language. In order to ensure the accuracy of data, the experiment under same flame resistance condition was repeated three times at least. And deviations of ΔP_{max} and T_{max} were less 5.60% and 7.90%, respectively.

Z. Wang and X. Cao, *Flame Resistance Characteristics and Its Influence Mechanism*,
https://doi.org/10.1007/978-981-95-6119-3_8

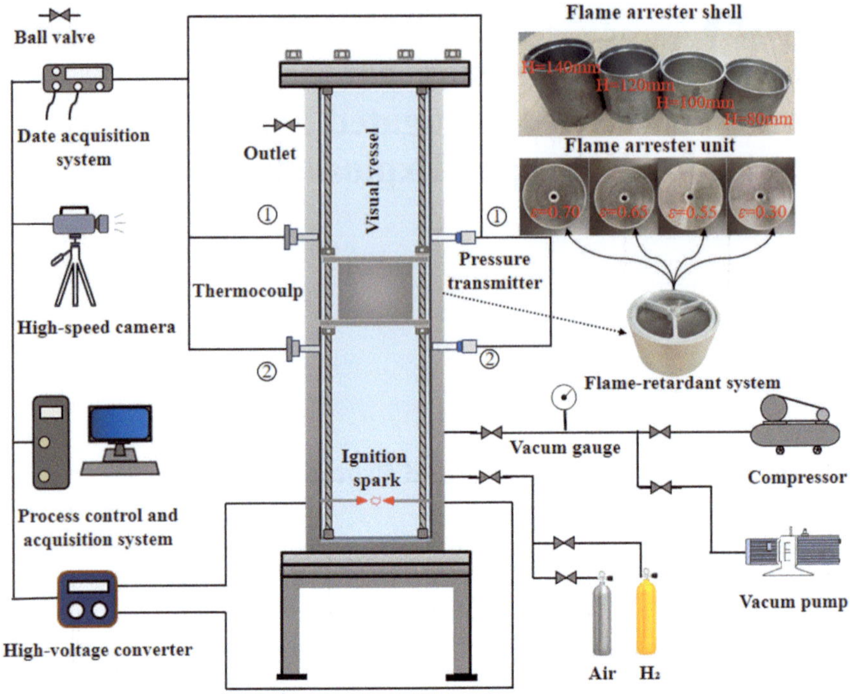

Fig. 8.1 Visualization hydrogen corrugated flame resistance apparatus

Table 8.1 Experimental condition parameters

	Hydrogen concentration (c)/%	Porosity (ε)	Flame arrester unit thickness (H)/mm
1	20	0.30	20
2	40	0.50	40
3	60	0.65	60
4	80	0.70	80

8.2 Results and Discussions

8.2.1 Analysis of Flame Resistance Characteristic

Figure 8.2 illustrates the relationship between flame propagation and pressure rise during a hydrogen explosion as flame resistance fails (c = 18%, ε = 0.30, H = 140 mm). There is a clear correlation between the flame passing through the corrugated flame arrester and the pressure rise under failure conditions. Initially, the flame developed a "spherical" structure. At 4.75 ms, the pressure at the lower end began to rise, but the slower flame propagation velocity resulted in a smaller pressure increase.

As the flame accelerated, it transitioned from an "ellipsoidal" to a "finger" structure, leading to a noticeable increase in the rate of pressure rise. Under the influence of a reverse shock wave, the velocity history exhibited significant oscillations. As the flame approached the lower end of the flame arrester, an axial acceleration process occurred. At 10.25 ms, the velocity history peaked. Subsequently, the flame entered the narrow channel and stagnated. At 11.75 ms, the flame passed through the narrow channel, and the pressure rise rate reached its peak at 14.50 ms. After this point, the pressure rise rate at the lower end began to decrease, with a pressure peak occurring at 17.50 ms ($\Delta P_{\max} = 0.32$ MPa). In comparison to the lower end, the moment of pressure rise at the upper end was slightly delayed (t = 6.75 ms). Following passage through the flame arrester, the pressure rise at the upper end was not significant; however, it was noticeably higher than that at the lower end in relation to flame propagation velocity. With continued acceleration of flame propagation, the rate of pressure rise increased markedly. At t = 14.00 ms, the velocity history at the upper end reached its peak; however, the corresponding pressure rise rate did not reach its maximum until later. As explosive reactions progressed at the upper end, heat accumulation increased continuously. At 18.25 ms, the pressure rise rate peaked and was significantly higher than that at the lower end. Subsequently, as explosive reactions weakened, the pressure rise rate decreased, with a pressure peak appearing at 19.50 ms ($\Delta P_{\max} = 0.36$ MPa). Compared to the lower end, $\Delta P_{\max}$ increased by 11.11%. Following this peak, pressure decreased significantly as the flame continued to extinguish.

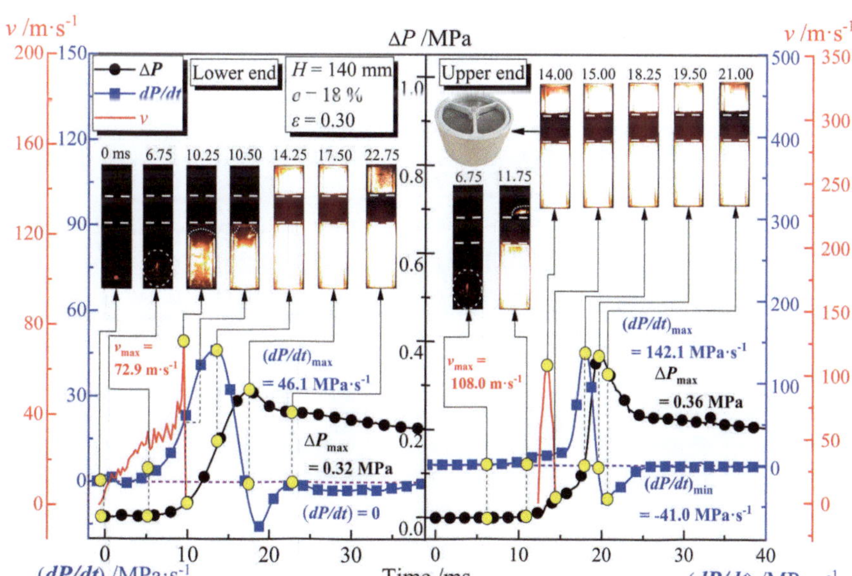

Fig. 8.2 Analysis of flame resistance characteristic under the failure condition ($c = 18\%$, $\varepsilon = 0.30$, $H = 140$ mm)

8.2.2 Analysis of Flame Propagation Process

Effect of Flame-Retardant Unit Thickness

Figure 8.3 illustrates the effect of flame-retardant unit thickness on the flame propagation process (c = 15%, ε = 0.30). By calculating the derivative of flame front height over time, the velocity history can be obtained (see in Fig. 8.4). It is evident that the initial flame evolves from a "spherical" to an "ellipsoidal" structures. As the thickness of the flame-retardant unit increases, the velocity history is significantly reduced, particularly during the acceleration phase of the "finger" flame. This reduction occurs because the reverse action of the high-speed shock wave is intensified due to the obstruction caused by the flame arrester, which continuously accumulates pressure at the front end and slows down flame propagation.

However, an evident acceleration phenomenon occurs as the flame approaches the end face of the flame-retardant arrester. At 16.00 ms, the flame first enters the narrow channel with a unit thickness of 80 mm. Subsequently, as the flame continues to enter the narrow channel, stagnation occurs, and the stagnation time gradually increases with greater flame-retardant unit thickness. At 22.00 ms, the flame passes through the narrow channel with H = 80 mm and accelerates, indicating a failure in flame resistance. In comparison to the lower end, the acceleration of flame propagation is more pronounced. As the thickness of the flame-retardant unit increases, the flame does not pass through the arrester to ignite the premixed gas at the upper end, demonstrating successful flame resistance. This indicates that variations in flame-retardant

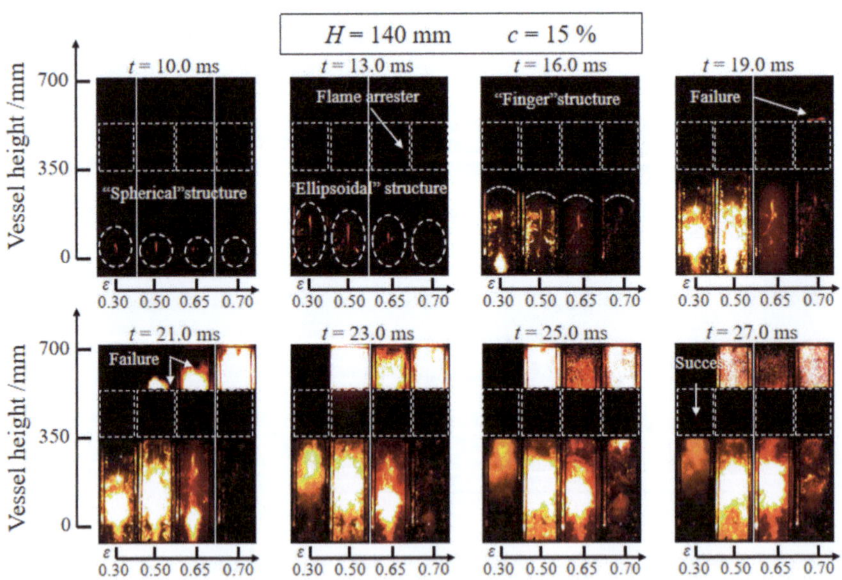

Fig. 8.3 Effect of flame-retardant unit thickness on flame propagation process ($c = 15\%, \varepsilon = 0.30$)

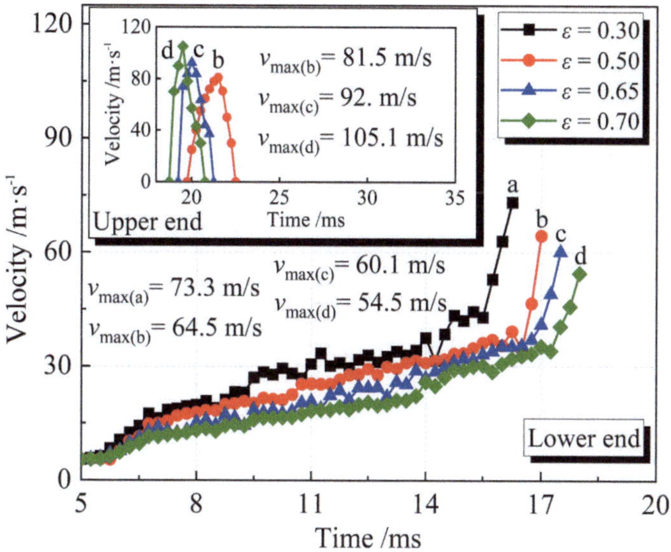

Fig. 8.4 Effect of flame-retardant unit thickness on velocity history ($c = 15\%$, $\varepsilon = 0.30$)

unit thickness can alter the velocity of flame propagation into the narrow channel, thereby affecting both shock wave attenuation and flame behavior within it.

Effect of Porosity

Figure 8.5 illustrates the effect of porosity on the flame propagation process ($c = 15\%$, $H = 140$ mm). As shown in the velocity history (in Fig. 8.6), the flame development rate progressively slows down as porosity increases, particularly during the development stages of "spherical" and "ellipsoidal" flames. This occurs because lower porosity enhances the reverse action of the shock wave, leading to a greater disturbance at the flame front and a higher flame propagation velocity. As porosity increases, the propagation velocity of the flame decreases sequentially as it enters the flame arrester, dropping from 73.3 to 54.5 m/s.

However, the stagnation time of the flame also decreases correspondingly. Notably, the flame first passes through the narrow channel and ignites the premixed gas at the upper end when $\varepsilon = 0.70$. Under a porosity condition of 0.30, the flame fails to pass through the narrow channel, resulting in successful flame resistance. Although a decrease in porosity can increase the flame propagation velocity at its front end, it also enhances the attenuation effects of the narrow channel on both the flame and shock wave. Furthermore, the increase in flame propagation velocity after passing through the flame arrester is more pronounced with higher porosity and decreases with lower porosity.

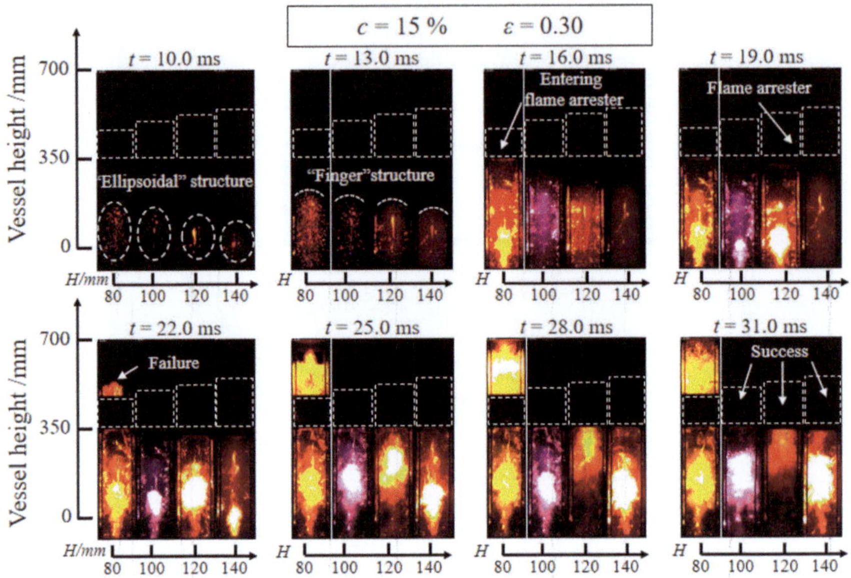

Fig. 8.5 Effect of porosity on flame propagation process ($c = 15\%$, $H = 140$ mm)

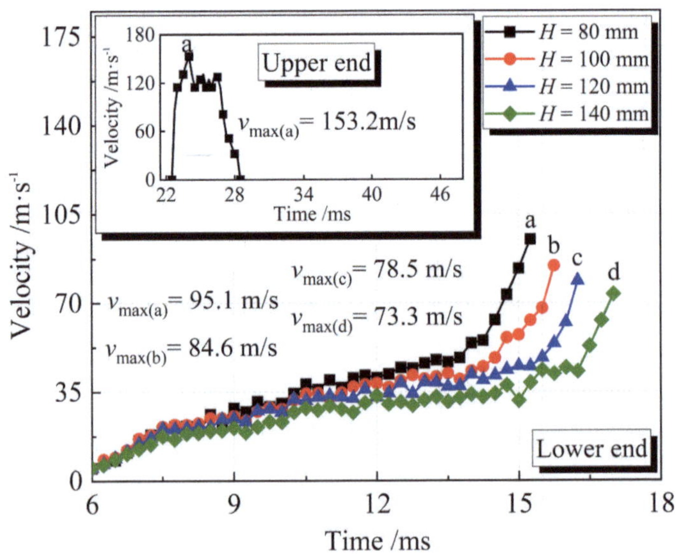

Fig. 8.6 Effect of porosity on velocity history ($c = 15\%$, $H = 140$ mm)

Effect of Hydrogen Concentration

Figure 8.7 illustrates the effect of hydrogen concentration on the flame propagation process ($\varepsilon = 0.30$, $H = 140$ mm). As shown in the velocity history in Fig. 8.8, the flame development rate gradually increases with higher hydrogen concentration, particularly during the development stages of "spherical" and "ellipsoidal" flames. This increase is attributed to the significant impact of hydrogen concentration on flame propagation velocity. Under lean-burn conditions, the velocity of flame propagation increases sequentially as hydrogen concentration increases, leading to reductions in both the moment of entry into the flame arrester and stagnation time within the narrow channel.

As $c = 21\%$, the flame first passes through the flame arrester and ignites the premixed gas at the upper end, indicating a failure in flame resistance. Meanwhile, the velocity after passing through the arrester gradually increases with higher hydrogen concentration. However, at $c = 12\%$ and 15%, the flame stagnates within the narrow channel, indicating successful flame resistance. This suggests that the propagation velocity of the flame entering the corrugated flame arrester is a key factor affecting flame resistance and significantly influences acceleration of flame propagation at the rear end under failure conditions.

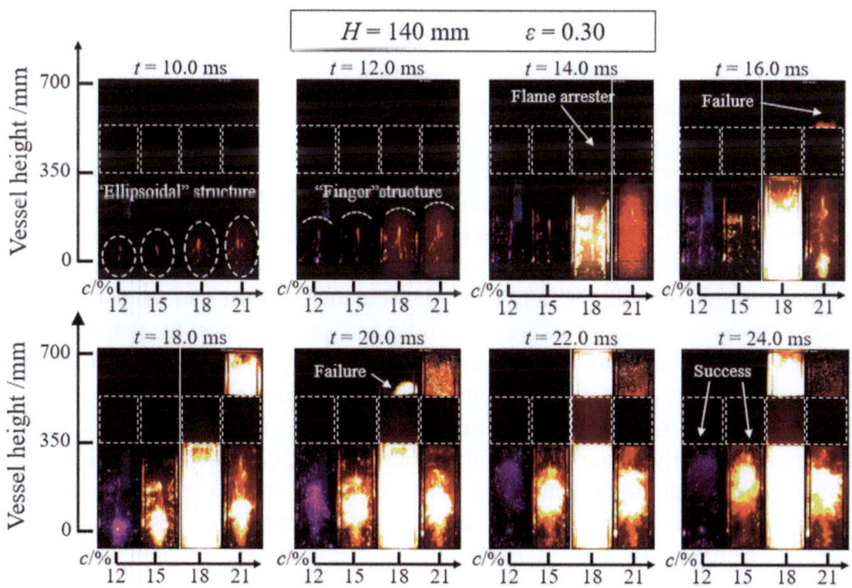

Fig. 8.7 Effect of hydrogen concentration on flame propagation process ($\varepsilon = 0.30$, $H = 140$ mm)

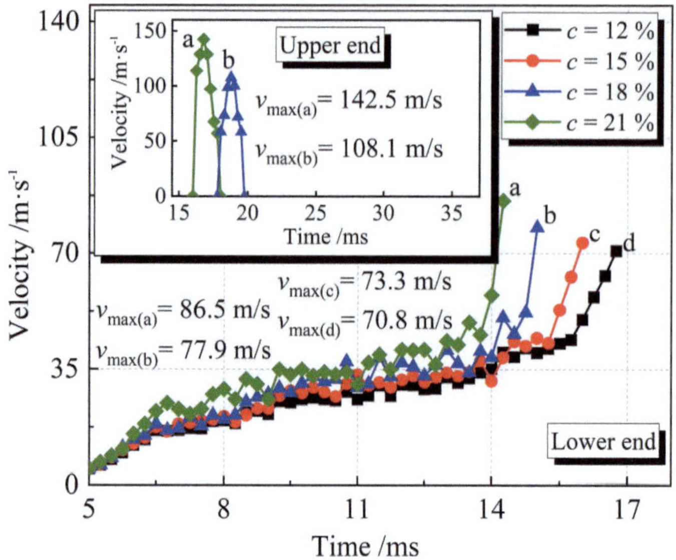

Fig. 8.8 Effect of hydrogen concentration on velocity history ($\varepsilon = 0.30$, $H = 140$ mm)

8.2.3 *Analysis of Explosion Pressure*

Effect of Flame-Retardant Unit Thickness

Figure 8.9 illustrates the effect of flame-retardant unit thickness on explosion pressure ($\varepsilon = 0.30$, $c = 15\%$). It is evident that the influence of flame-retardant unit thickness on pressures at both ends aligns with its impact on velocity history. When $H = 80$ mm, flame resistance fails, resulting in the upper pressure being significantly higher than the lower pressure, with an increase of 25.71% at the upper end compared to the lower end. As the thickness of the flame-retardant unit increases, the flame does not pass through the flame arrester; consequently, the lower pressure becomes noticeably higher than the upper pressure, exhibiting a slight decreasing trend. Compared to the failure condition ($H = 80$ mm), pressure peaks at both ends decrease by 60.0% and 7.70%, respectively, under successful flame resistance conditions ($H = 100$ mm). This indicates that the pressure drop at the upper end is more pronounced during successful flame resistance; however, the extent of reduction becomes less significant as flame-retardant unit thickness continues to increase. As H = 140 mm, the pressure peaks at both ends decrease by 20.83% and 14.29%, respectively. The history of pressure rise rate can be obtained by calculating the derivative of pressure history over time. As the thickness of the flame-retardant unit increases, changes in the pressure rise rate exhibit a pattern similar to that of the pressure.

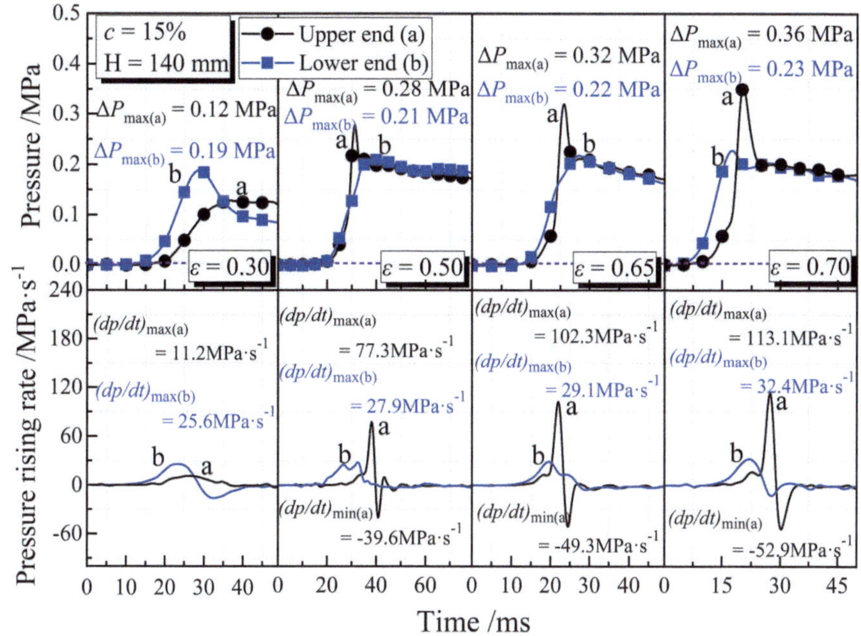

Fig. 8.9 Effect of flame-retardant unit thickness on pressure history ($\varepsilon = 0.30$, $c = 15\%$)

Effect of Porosity

Figure 8.10 illustrates the effect of porosity on explosion pressure ($H = 140$ mm, $c = 15\%$). It is evident that both the pressure history and its rising rate exhibit similar change patterns with increasing porosity. At $\varepsilon = 0.30$, flame resistance is successful, with lower pressure exceeding upper pressure. However, as porosity increases to 0.50, the flame resistance fails, resulting in a significant increase in upper pressure and its rising rate. The pressure peaks at both ends increase by 57.14% and 9.52%, respectively. Notably, the change in the pressure rising rate is more pronounced than that of the pressure peak. As porosity increases, upper pressure rises, and the corresponding pressure rising rate history follows a similar variation pattern. ΔP_{max} at upper and lower ends are increased by 66.67% and 17.39%, respectively. And $(dP/dt)_{max}$ are increased by 90.10% and 20.99%, respectively. This indicates that an increase in porosity during flame resistance failure generates greater pressure at the rear end of the flame arrester. Therefore, the structural parameters of corrugated flame arresters selected for engineering applications must ensure effective flame resistance; otherwise, they may enhance pressure effects and cause serious damage. Considering allowable pressure drops, further optimization of flame-retardant system structure parameters is necessary.

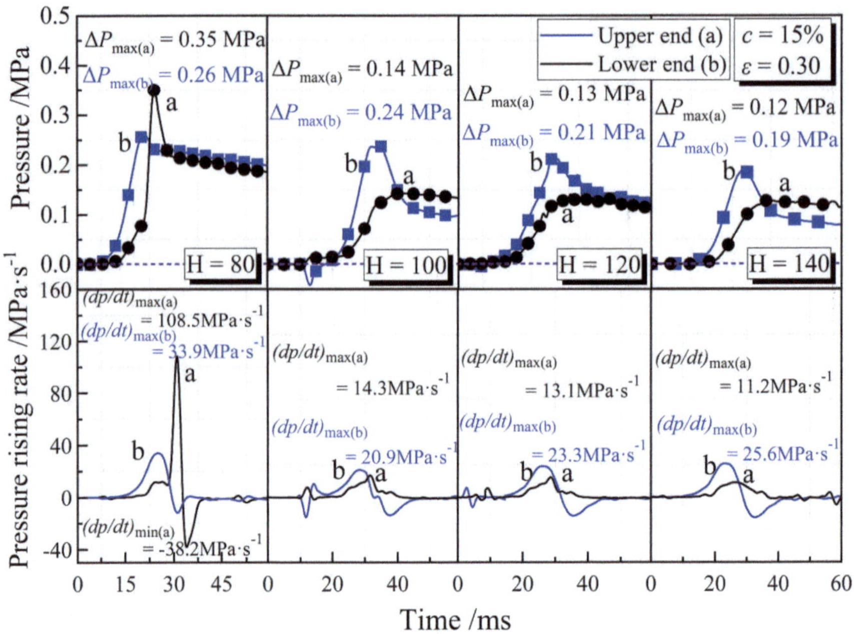

Fig. 8.10 Effect of porosity on pressure history (H = 140 mm, c = 15%)

Effect of Hydrogen Concentration

Figure 8.11 illustrates the effect of hydrogen concentration on explosion pressure (H = 140 mm, ε = 0.30). Hydrogen concentration significantly impacts explosion pressure and its rising rate at both ends, influencing the explosion reaction rate, which in turn affects the propagation velocity and intensity of the shock wave entering the narrow channel. The relationship between hydrogen concentration and explosion pressures, as well as their rising rates at both ends, follows a similar variation pattern; however, pressure changes at the two ends are not significant under opposing flame resistance effects. At c = 12% and 15%, flame resistance is successful, with lower pressure exceeding upper pressure. Conversely, at c = 18% and 21%, the flame resistance fails, resulting in upper pressure being significantly higher than lower pressure. As hydrogen concentration increases, pressures at both ends exhibit a slight upward trend. This indicates that hydrogen concentration has a notable impact on the rates of pressure rise at both the front and rear ends of the flame-retardant system, with the likelihood of failure increasing as concentration rises.

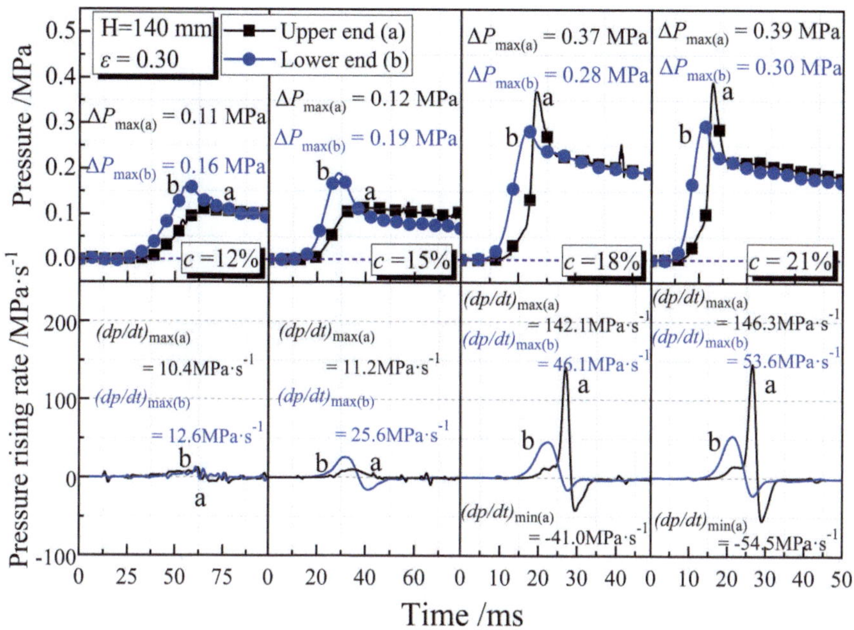

Fig. 8.11 Effect of hydrogen concentration on pressure history ($H = 140$ mm, $\varepsilon = 0.30$)

8.2.4 *Analysis of Flame Temperature*

Effect of Flame-Retardant Unit Thickness

Figure 8.12 shows the effect of flame-retardant unit thickness on the flame temperature peak ($\varepsilon = 0.30$, $c = 15\%$). It clear that the change in flame temperature peak (T_{max}) presented the same pattern as the change in pressure peak. When $H = 80$ mm, the flame resistance fails, resulting in the temperature at the upper end being higher than that at the lower end. As the thickness of the flame-retardant unit increases, the flame resistance is successful, and T_{max} decreases significantly, particularly at the upper end where the drop is more pronounced. Compared to a unit thickness of 80 mm, T_{max} at the upper and lower ends corresponding to a unit thickness of 100 mm decreases by 79.48% and 15.63%, respectively. Meanwhile, T_{max} at the lower end remains noticeably higher than at the upper end and does not reach the ignition temperature of hydrogen. Although the flame does not pass through the flame-retardant system, it still exhibits a significant temperature increase due to high-temperature unburned gas. This indicates that as unit thickness increases, the contact time between the narrow channel and the flame also increases, further enhancing heat absorption by the flame and consequently reducing the temperature at the flame front.

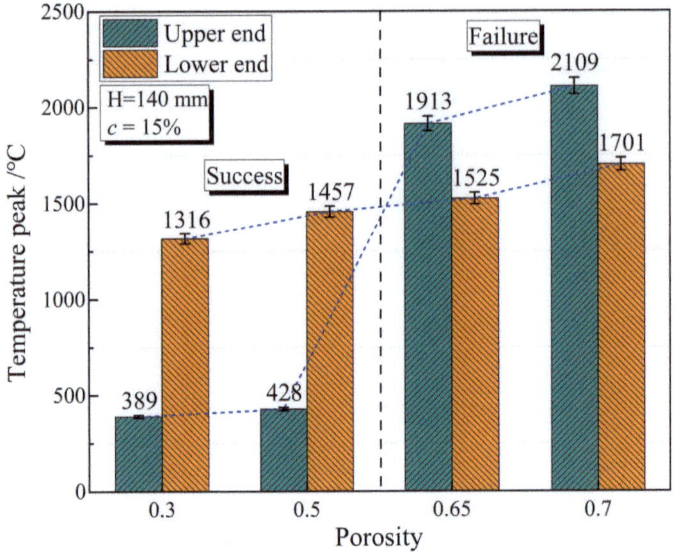

Fig. 8.12 Effect of flame-retardant unit thickness on temperature history ($\varepsilon = 0.30$, $c = 15\%$)

Effect of Porosity

Figure 8.13 presents the effect of porosity on the flame temperature peak ($c = 15\%$, $H = 140$ mm). It is evident that T_{max} at both ends under successful flame resistance is significantly lower than the failure, with the change in T_{max} at the upper end being more pronounced. When flame resistance is successful, T_{max} at the upper end is lower than at the lower end. Compared to a porosity of 0.30, T_{max} at the upper and lower ends corresponding to a porosity of 0.50 increases by 9.11% and 9.68%, respectively. However, when flame resistance fails, T_{max} at the upper end exceeds that at the lower end. Compared to a porosity of 0.45, T_{max} at the upper and lower ends corresponding to a porosity of 0.65 increases by 9.29% and 10.35%, respectively. Additionally, T_{max} at both ends corresponding to a porosity of 0.45 increases by 77.63% and 4.46%, respectively, compared to a porosity of 0.25. This indicates that reducing porosity increases the surface area of the flame in contact with the inner wall of the narrow channel, thereby enhancing the heat transfer rate and wall effect.

Effect of Hydrogen Concentration

Figure 8.14 shows the effect of hydrogen concentration on the flame temperature peak ($\varepsilon = 0.30$, $H = 140$ mm). It is evident that hydrogen concentration significantly affects the flame temperatures at both the upper and lower ends, thereby influencing the flame resistance effectiveness of the corrugated flame arrester in hydrogen explosions. At $c = 12\%$ and 15%, flame resistance is successful, with T_{max} at the lower end

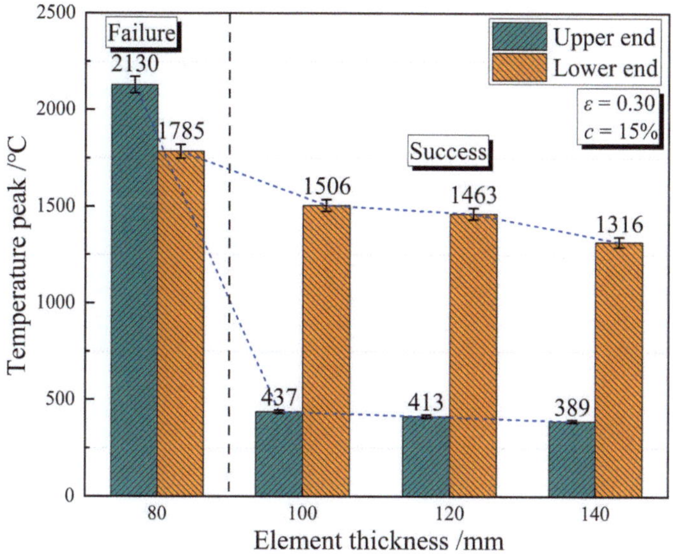

Fig. 8.13 Effect of porosity on temperature history ($c = 15\%$, $H = 140$ mm)

exceeding that at the upper end. Additionally, T_{max} increases noticeably with rising hydrogen concentration. Compared to a hydrogen concentration of 12%, T_{max} at the upper and lower ends corresponding to a concentration of 15% increases by 21.34% and 14.29%, respectively. However, at $c = 18\%$ and 21%, the flame resistance fails, resulting in T_{max} at the upper end being higher than at the lower end. As hydrogen concentration increases, T_{max} sequentially rises. Compared to a hydrogen concentration of 18%, T_{max} at the upper and lower ends corresponding to a concentration of 21% increases by 12.33% and 10.28%, respectively. This indicates that an increase in hydrogen concentration enhances both flame propagation velocity and temperature as they enter the narrow channel of the flame-retardant system, thus affecting the attenuation effect of the narrow channel on the flame. Furthermore, an increase in hydrogen concentration is detrimental to the explosion resistance of the corrugated flame arrester.

8.2.5 Analysis of Flame Resistance Mechanism

Figure 8.15 illustrates the schematic diagram of the hydrogen explosion flame quenching process within the narrow channel of a corrugated flame arrester. The flame resistance mechanism can be attributed to a combination of physical and chemical effects [2]. In terms of physical effects, the high-speed shock wave acted on the end face of flame-retardant unit before the high-temperature flame. Firstly, part of shock waves emerged the reverse propagation due to the obstruction of end face,

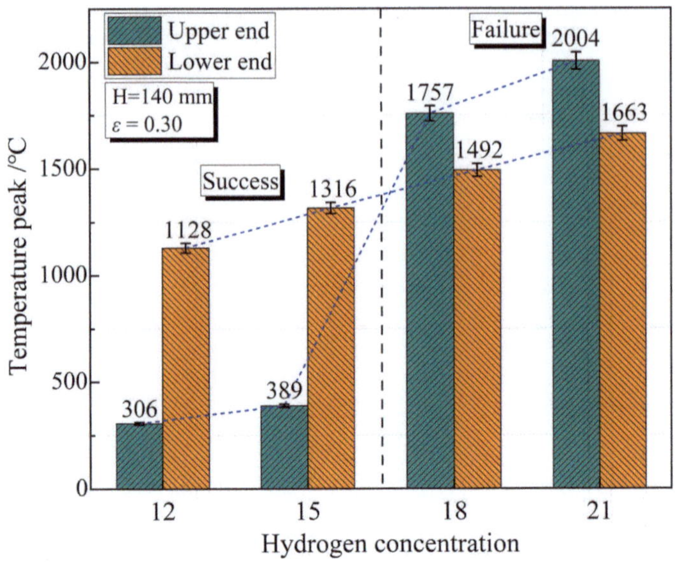

Fig. 8.14 Effect of hydrogen concentration on temperature history ($\varepsilon = 0.30$, $H = 140$ mm)

thereby weakening the intensity of shock wave [3]. The reverse shock wave also exerts a blocking effect on flame propagation, reducing its velocity as it enters the narrow channel [4]. Meanwhile, discrete shock waves enter the narrow channel and continuously collide with the wall surface during propagation, further diminishing their intensity. As the flame came into contact with the end face of flame-retardant system, the continuous and smooth flame was also dispersed by the narrow channel and propagated inside it. The flame continuously exchanged heat with the wall face during propagation, which greatly reduced the temperature of flame front. As the temperature dropped below the ignition temperature of hydrogen, the flame resistance was successful [5]. Apart from the physical effect, the number and energy of free radicals activated for the main chain reactions ($H_2 + O_2 = HO_2 + H$, $H + O_2 = O + OH$, $OH + H_2 = H_2O + H$) of high-temperature flame caused by the hydrogen explosion were greatly reduced during the interaction with the wall surface. As the production rate of chain reaction was less than the destruction rate, the flame was quenched. The thickness and porosity of the flame-retardant unit, along with hydrogen concentration, influence flame resistance by altering shock wave intensity and flame behavior as they enter the narrow channel, as well as affecting physical and chemical interactions during internal propagation.

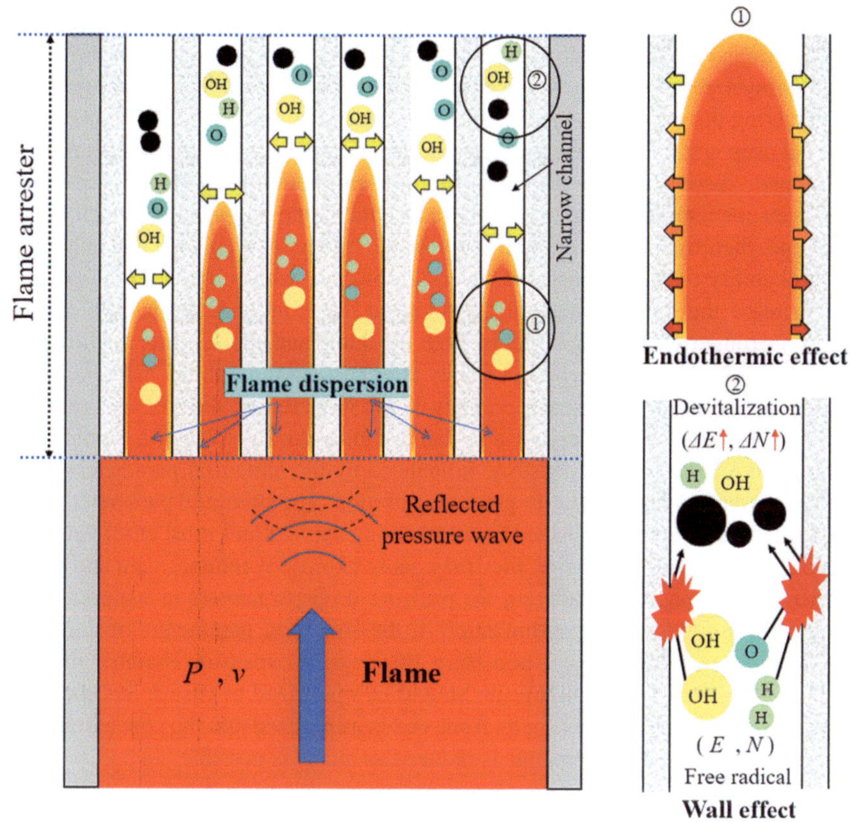

Fig. 8.15 Schematic diagram of flame resistance mechanism

8.3 Conclusions

This experiment primarily investigated the flame resistance characteristics and influence mechanisms of the corrugated flame-retardant system on hydrogen explosions. Through visual flame resistance experiments, we clarified the effects of the flame-retardant system structure parameters and hydrogen concentration on flame propagation characteristics and explosion parameters at both ends. Additionally, the influences of flame propagation and shock wave attenuation within the narrow channel were summarized, further exploring the underlying explosion resistance mechanisms in depth. The conclusions are as follows:

First, a clear correlation exists between flame propagation and pressure rise at both ends of the flame-retardant system during explosion resistance. An evident oscillation occurs in the initial stage of flame propagation due to obstruction by the end face of the flame-retardant system. Additionally, the flame-retardant system significantly influences the flame propagation process at its front end, thereby affecting its velocity within the narrow channel. As flame resistance fails, the flame at the rear end experiences a noticeable acceleration. The structural parameters of the flame-retardant system and hydrogen concentration influence the velocity of flame propagation entering the narrow channel, as well as the physical and chemical effects during internal propagation, ultimately impacting the effectiveness of flame resistance.

Second, increasing the thickness of the flame-retardant unit and decreasing hydrogen concentration can effectively reduce the flame propagation velocity at the front end, thereby influencing its velocity within the narrow channel and enhancing the probability of flame resistance success. However, while a reduction in porosity can increase the flame propagation velocity at the front end, the narrow channel exerts a stronger weakening effect on both the flame and shock wave. Additionally, the increase in flame propagation velocity after passing through the narrow channel is more pronounced with greater porosity, while it decreases with reduced porosity.

Third, the explosion pressure and flame temperature displayed a similar variation pattern to the velocity history, influenced by the structural parameters of the flame-retardant system and hydrogen concentration. The structural parameters of the corrugated flame arrester selected for engineering applications must ensure effective explosion resistance; otherwise, they may cause explosion enhancement and lead to catastrophic damage. Considering the pressure drop parameters in a normal operating environment, further optimization of the structural parameters of the corrugated flame-retardant system is necessary. Besides, the expansion chamber structure at the front end of the flame arrester should be optimized to ensure that the flame propagation velocity of entering its front-end is minimized and the concentration of transported gas inside the pipeline is reduced as much as possible.

References

1. Cao XY, Zhou X, Huang R, Wang Z, Wang ZR. Research on the resistance characteristics of corrugated flame-retardant system for hydrogen/methane explosion. Int J Hydrogen Energy. 2024;88:228–41.
2. Lin CD, Cao XY, Wang ZR, et al. Research on quenching performance and multi-factor influence law of hydrogen crimped-ribbon flame arrester using response surface methodology. Fuel. 2022;326:124911.
3. Mahuthannan AM, Damazo JS, Kwon E, et al. Effect of propagation speed on the quenching of methane, propane and ethylene premixed flames between parallel flat plates. Fuel. 2019;256:115870.
4. Bao L, Wang P, Dang WY, et al. Experimental study on detonation flame penetrating through flame arrester. J Loss Prev Process Ind. 2021;72:104529.
5. Wang LQ, Ma HH, Shen ZW. The quenching of propane deflagrations by crimped ribbon flame arrestors. J Loss Prev Process Ind. 2016;43:567–74.

Chapter 9
Research on the Resistance Characteristics of Corrugated Flame-Retardant System for Hydrogen–Methane Explosion

9.1 Experimental Apparatus and Procedures

The experimental apparatus of corrugated flame-retardant system was shown in Fig. 9.1. The main part had been detailed in the previous work. Only a detailed explanation was introduced for the corrugated flame-retardant system. It consisted of the flame-retardant element shell, the flame-retardant element, the flame-retardant pressure plate, and four threaded support rods. The flame-retardant system was made of 304 stainless steel and its thermal conductivity was 16.3 W m^{-1} K^{-1}. Multilayer flame-retardant elements were tightly stacked inside the flame-retardant element shell and was fastened by bolts, and placed inside the upper and lower flame-retardant pressure plates. The flame-retardant pressure plates with four through-holes at the diagonals were installed on the support rod with a determined height ($H = 650$ mm). By changing the length of flame-retardant element shell and the number of flame-retardant element, the structural parameters for different corrugated flame-retardant systems were achieved. To ensure a tight joint between the flame-retardant system and the inner wall surface, the gaps was sealed by the high temperature explosion-proof sealing mud. Two visualization windows were installed at the front and rear ends of explosion vessel to capture the transient transformation process of flame structure obstructed by the flame-retardant system and the propagation change after passing through the narrow channel.

During the experiment, a high-speed camera [Acquisition frequency (f) = 4000 fps] was used to record the dynamic flame evolution characteristics during the explosion resistance. The coordinate position of the flame front was determined through the software analysis system, and the height change of the flame front within 0.25 ms was obtained through proportional conversion. Two high frequency dynamic pressure transducers with a range of 0–2.5 MPa and a response frequency of 50 kHz were used to collect pressure data at two ends of flame-retardant system. Two S-type platinum–rhodium thermocouple [Range (A) = 3000 °C] were selected to collect flame temperature data. The ignition and data acquisition were controlled by self-written

Z. Wang and X. Cao, *Flame Resistance Characteristics and Its Influence Mechanism*, https://doi.org/10.1007/978-981-95-6119-3_9

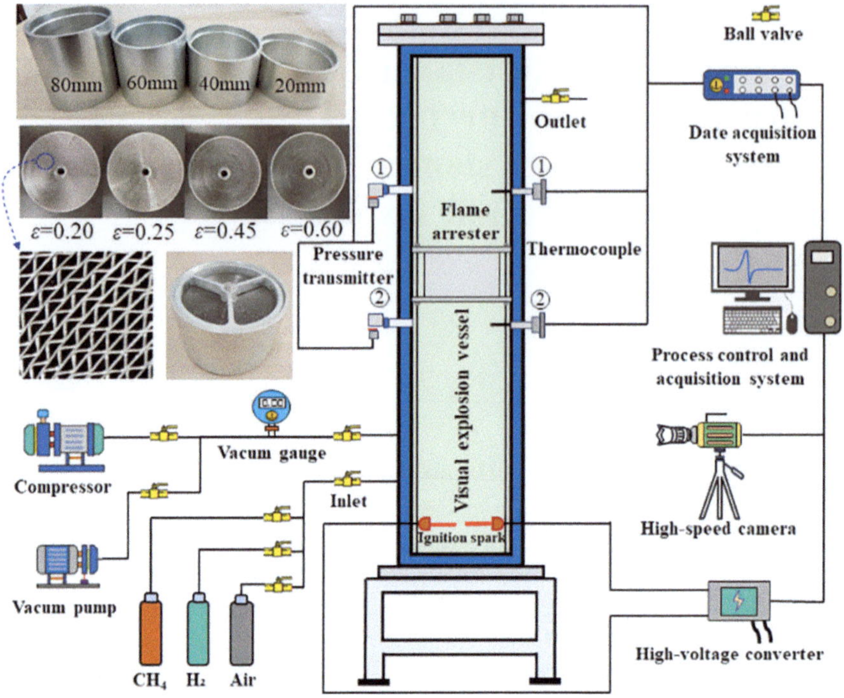

Fig. 9.1 Schematic diagram of experimental setup

language and the program-control acquisition system [Acquisition frequency (f) = 200 kHz/channel].

The structural components of corrugated flame-retardant system were all self-processed. The thickness of flame-resistant element was 20 mm and its diameter was 110 mm. Four types of flame-resistant element thicknesses (H = 20 mm, 40 mm, 60 mm and 80 mm) and four types of porosities ($\varepsilon = 0.30$, 0.50, 0.65 and 0.70) were selected as its structural parameters during the experiment. According to Dalton's law of partial pressure, the hydrogen doped methane premixed gas with four kinds of hydrogen ratios [(Φ) = 20%, 40%, 60% and 80%] at different equivalence ratios ($\varphi = 0.4$, 0.6, 0.8 and 1.0) were prepared. The experiment condition was shown in Table 9.1. Based on the correspondence between flame propagation and parameters during the explosion resistance of corrugated flame-retardant system, as well as the influence of working conditions on the explosion resistance parameter, the influence law and its mechanism were revealed. It made up the experimental gap in the visualization research of corrugated flame-retardant system to prevent the high actively flammable gas explosion. To ensure the accuracy of data, the experiment was repeated three times under the same operating condition. Through data analysis, the maximum errors of pressure peak (ΔP_{max}) and temperature peak (T_{max}) did not exceed 4.0% and 3.1%, respectively.

Table 9.1 Detailed information of experimental conditions

	Equivalence ratio (φ)	Hydrogen ratio (Φ)/%	Porosity (ε)	Element thickness (H)/mm
1	0.4	20	0.30	20
2	0.6	40	0.50	40
3	0.8	60	0.65	60
4	1.0	80	0.70	80

9.2 Results and Discussions

9.2.1 Analysis of Explosion Resistance Characteristics

To clarify the explosion resistance characteristics of corrugated flame-retardant system on hydrogen doped methane, the correspondence between the flame structure transformations and pressure changes at the front and rear ends of flame-retardant system was explored. Figure 9.2 shows the corresponding relationship between the two over time under the flame resistance failure (H = 20 mm, φ = 0.6, Φ = 80%, ε = 0.70). Under this working condition, the flame passed through the flame-retardant element. After igniting, the flame developed in a "spherical" structure, and there was no significant change in the pressure histories at the two ends. As t = 6.00 ms, the lower pressure history began to rise. And the flame changed from the "spherical" to "finger" structure and accelerated upward propagation at 10.00 ms. As the flame accelerated, the rate of pressure rise (*dP/dt*) at the lower end was obviously increased. The reverse propagation of shock wave caused by the obstruction of flame resistance system end surface resulted in an obvious oscillation of velocity history. As t = 17.75 ms, the flame approached the flame-retardant system and exhibited an axial acceleration change, and the velocity history reached its peak at this moment (v_{max} = 47.15 m/s). And it entered the flame-retardant system at 18.50 ms and appeared stagnation. After passing through, (*dP/dt*) history at the lower end was slightly decreased. Afterwards, the explosion reaction intensified and (*dP/dt*) increased again due to the continuous accumulation of heat at the lower end. As t = 26.50 ms, (*dP/dt*) history reached its peak. Subsequently, the lower flame was gradually extinguished and (*dP/dt*) continued to decrease. At 32.00 ms, the lower pressure history reached its peak. For the upper end, the flame stagnated inside the flame-retardant system for 2.50 ms before passing through the rear end surface and igniting the upper premixed gas at 21.00 ms. Because of narrow channel, the discrete flame accelerated propagation and reached the velocity history peak at 22.00 ms (v_{max} = 159.20 m/s). With the acceleration propagation of flame and the continuous accumulation of heat, (*dP/dt*) was greatly increased and reached its peak at 24.00 ms. Afterwards, the reverse shock wave passed through the narrow channel, causing a sharp decrease in pressure rise rate and appearing an obvious negative value [$(dP/dt)_{min}$ = − 67.43 MPa/s]. Besides, ΔP_{max} of upper end had an obvious increase and appeared earlier compared to the lower end (t = 25.50 ms, ΔP_{max} = 0.44 MPa). This indicates that an enhancement

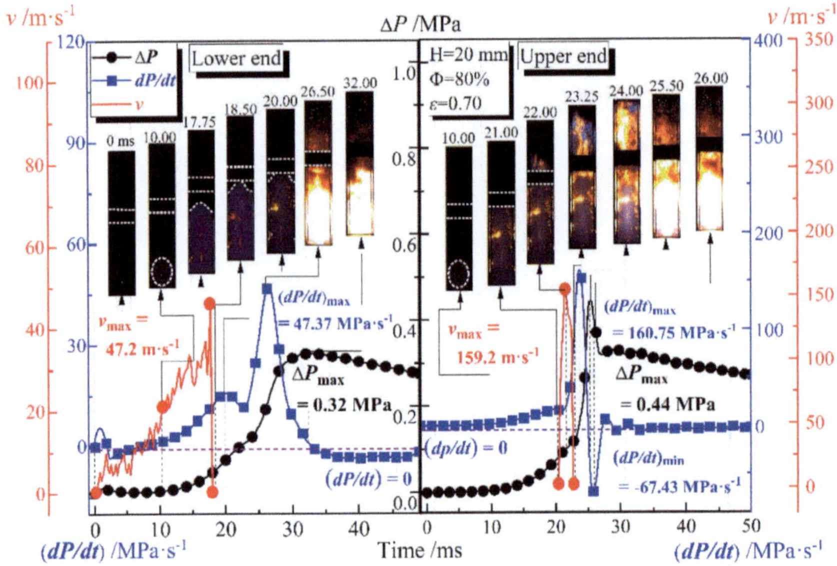

Fig. 9.2 Analysis of flame propagation characteristics under failure condition (H = 20 mm, φ = 0.6, Φ = 80%, ε = 0.70)

effect on its rear end would appear and lead to an increase in explosion parameters after the flame resistance failed.

9.2.2 Flame Propagation Characteristics

Effect of Flame-Retardant Element Thickness

Figure 9.3 shows the effect of flame-retardant element thickness on the flame propagation process (φ = 0.6, Φ = 80%, ε = 0.65). It clear that the initial flame developed in a "spherical" structure. Nevertheless, the enhancement of reflection pressure wave resulted in a slowdown of flame propagation velocity as the flame-retardant element thickness increased. Especially the impact was more significant during the acceleration propagation stage of "finger" flame, and the overall presented the light blue. As H = 20 mm, the flame first entered the flame-retardant system, and the front structure shrunk and accelerated propagation. Afterwards, the flame sequentially entered the corrugated flame-retardant system. Due to the continuous improvements of heat absorption efficiency and wall effect of narrow channel, its stagnation time inside the flame-retardant system was evidently prolonged. Based on the corresponding velocity history (seen in Fig. 9.4), it can be inferred that the peak value of lower velocity history was sequentially decreased and the time for the flame to pass through the

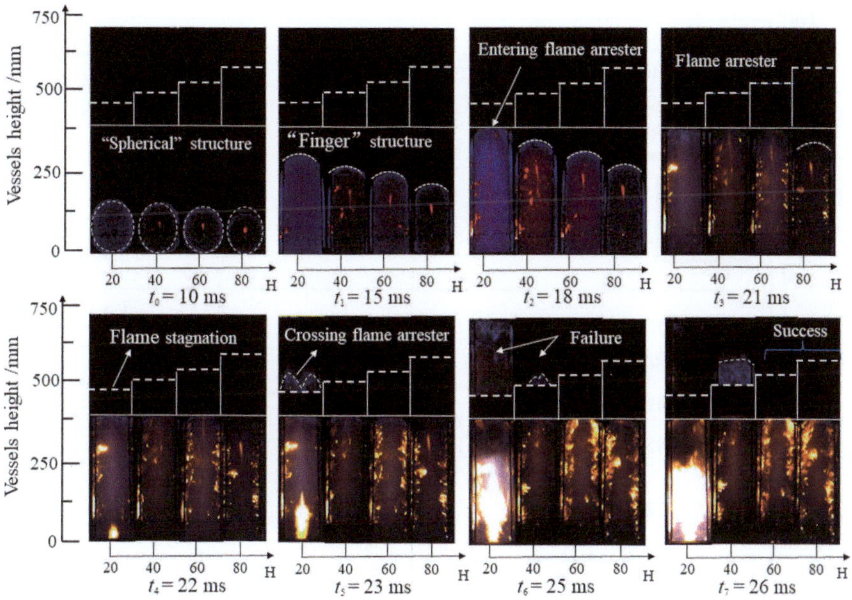

Fig. 9.3 Effect of flame-retardant element thickness on the flame propagation process (φ = 0.6, Φ = 80%, ε = 0.65)

flame-retardant system was prolonged as its element thickness increased [1]. Especially the flame did not pass through the narrow channel to ignite the above premixed gas as H = 60 mm. However, the flame passed through the flame-retardant element and presented an obvious acceleration phenomenon under the failure condition, and the acceleration effect became more significant with the decrease of element thickness. Through multi-condition experimental analysis, it was found that the effect of flame-retardant element thickness showed the similar patterns of change. This indicates that the corrugated flame-retardant element should have a critical thickness to effectively prevent flame from passing through, otherwise it would result in the enhancement of explosion.

Effect of Porosity

Figure 9.5 illustrates the effect of porosity on the velocity history (φ = 0.6, Φ = 80%, H = 20 mm). It clear that the disturbance effect of reflection shock wave on the flame front was enhanced as the porosity decreased. The velocity history of initial flame showed a slight increase trend. Especially it manifested in an obvious increase of $v_{\max}$ at the lower end and the advance timing of entering the flame-retardant system. However, the efficiency of heat transfers and the wall effect of narrow channel surface on flame continued to increase as the porosity decreased. It was manifested as a significant extension of stagnation time inside the flame-retardant

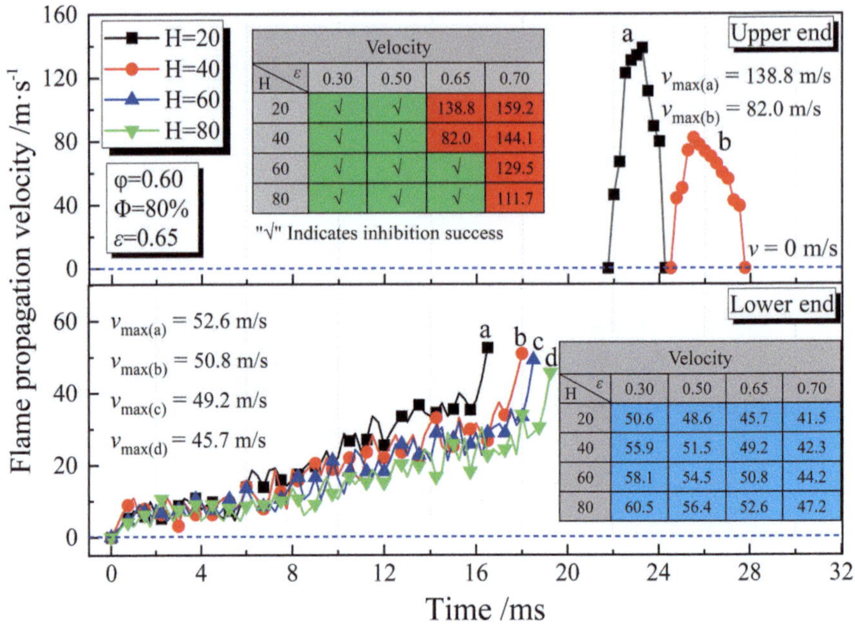

Fig. 9.4 Effect of flame-retardant element thickness on the velocity history (φ = 0.6, Φ = 80%, ε = 0.65)

system. Especially, the flame did not pass through the flame-retardant system to ignite the above premixed gas as ε = 0.50. This indicates that the decrease of porosity could effectively enhance the weakening effect of narrow channel on the flame propagation. Nevertheless, it would also cause obvious flame acceleration under resistance failure and affected by porosity. As the porosity and hydrogen ratio increased, the probability of flame-retardant failure was increased.

Effect of Hydrogen Ratio

Figure 9.6 shows the effect of hydrogen ratio on the velocity history (φ = 0.6, H = 20 mm, ε = 0.70). It clear that the hydrogen ratio had a more significant impact on the flame propagation compared to the structural parameter of flame-retardant system. In the initial stage, the rise slope of velocity history was evidently increased as the hydrogen ratio enhanced. Especially the velocity entering the flame-retardant system was greatly increased from 18.8 to 47.2 m/s. Under this operating condition, the flame first entered the flame-retardant system and presented the shortest stagnation time, then passed through it to ignite the upper premixed gas.

As the flame retardant failed, the narrow channel as an obstacle had a significant acceleration effect on flame propagation. Nevertheless, the velocity history was rapidly decreased, and its drop extent was even more evidently after passing through

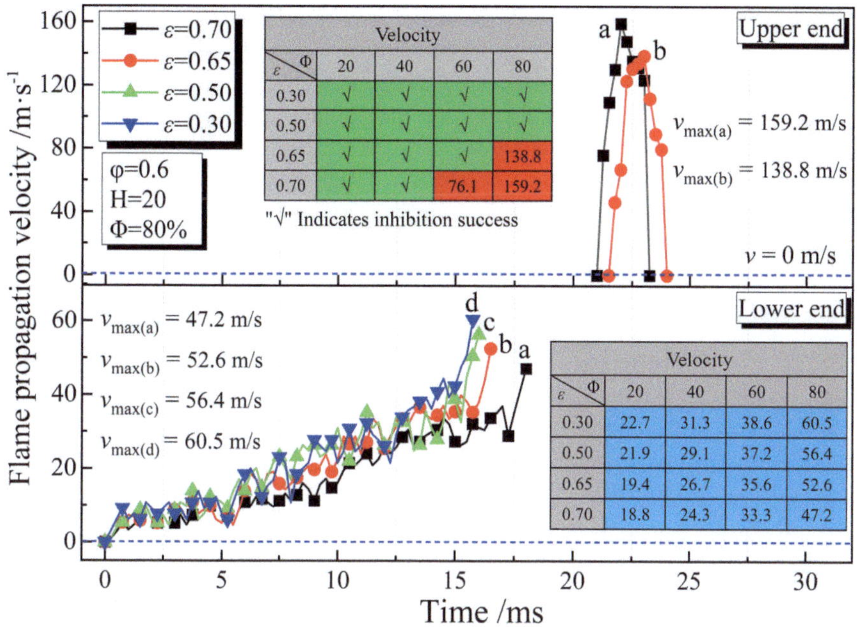

Fig. 9.5 Effect of porosity on the velocity history (φ = 0.6, Φ = 80%, H = 20 mm)

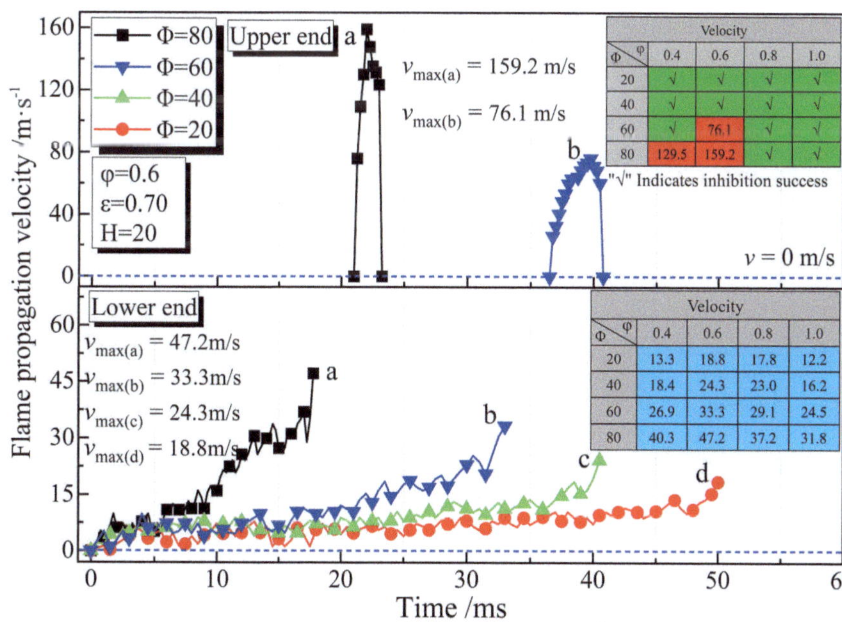

Fig. 9.6 Effect of hydrogen ratio on velocity history (φ = 0.6, H = 20 mm, ε = 0.70)

the flame-retardant system as the hydrogen ratio reduced. Especially the flame did not pass through the flame-retardant system as $\Phi = 40\%$, indicating flame resistance success. This indicates that an increase in hydrogen ratio could improve the propagation velocity of flame after the mixture explosion, which would also propose a higher demand on the flame-retardant performance. Besides, the flame resistance effect was related to its propagation velocity entering front end surface and affected by the comprehensive action of structural parameter.

9.2.3 *Analysis of Explosion Pressure*

Effect of Flame-Retardant Element Thickness

Figure 9.7 presents the effect of flame-retardant element thickness on the explosion pressure of two ends ($\varphi = 0.6$, $\Phi = 80\%$, $\varepsilon = 0.65$). It clear that the upper pressure was obviously higher than that of lower end as the flame resistance failed, and much higher than the pressure under the flame resistance success. This indicates that increase of flame-retardant element thickness could effectively reduce the explosion intensity of hydrogen doped methane premixed gas. However, the lower pressure was slightly higher than that of upper end under the success condition. With the increase of flame-retardant element thickness (from 20 to 40 mm), the pressures at the upper and lower end was decreased from 0.41 MPa to 0.36 MPa and from 0.31 MPa to 0.24 MPa, respectively. Especially, the pressure sharply dropped as the flame resistance was successful under 60 mm element thickness. Compared to 40 mm element thickness, ΔP_{max} at the upper and lower ends were decreased by 64.0% and 43.0%, respectively. It was because that the increase in element thickness extended the contact time of the flame and shock wave with narrow channel, greatly weakening the intensities of both. Ultimately, the spread of flame was effectively prevented and the quenching occurred, thereby increasing the probability of flame resistance success. Besides, an increase in the element thickness could slightly enhance the lower pressure under the flame resistance success.

Effect of Porosity

Figure 9.8 illustrates the effect of porosity on the pressure history ($\varphi = 0.6$, $\Phi = 80\%$, $H = 20$ mm). The decrease in the porosity could effectively improve the flame resistance effect and reduce the pressure. As the flame resistance failed, the pressures at both ends were evidently higher than the pressure under the success condition. As the porosity decreased, the pressure was gradually reduced. As $\varepsilon = 0.50$, the flame happened quenching inside the narrow channel and the pressure was sharply decreased. Compared with 0.70 porosity, ΔP_{max} at the upper and lower ends were decreased by 68.18% and 56.25%, respectively. It was because that the decrease in porosity led to the increases of end surface narrow channel number and shock wave

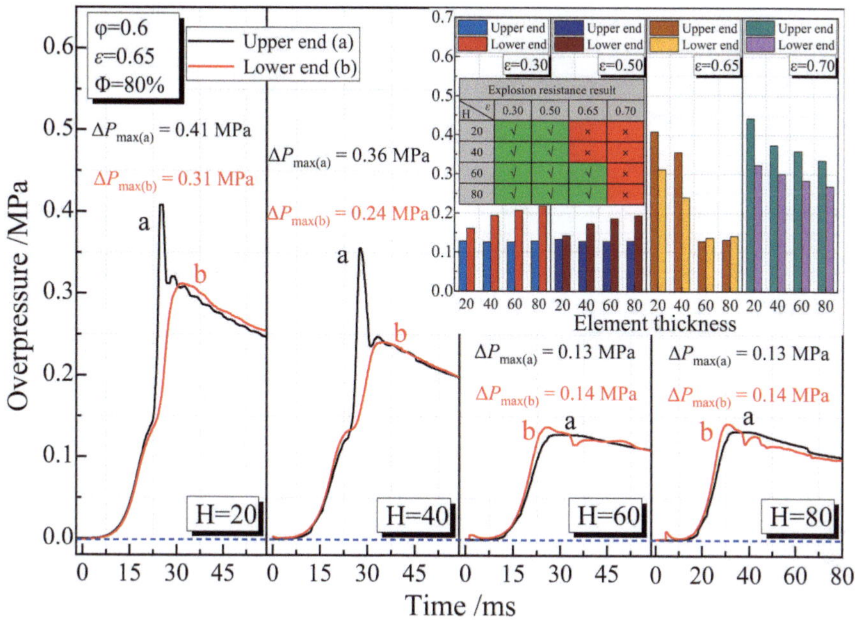

Fig. 9.7 Effect of flame-retardant element thickness on the pressure history (φ = 0.6, Φ = 80%, ε = 0.65)

dispersion extent, further resulting in an increasing weakening effect of shock wave during the propagation inside the narrow channel. Besides, there was no significant change in pressure as the flame was successful under four different hydrogen ratios. As the flame resistance failed, the pressure was greatly increased. Especially for the upper pressure. As the porosity and hydrogen ratio increased, the failure probability continued to increase.

Effect of Hydrogen Ratio

Figure 9.9 shows the effect of hydrogen ratio on the pressure history (φ = 0.6, ε = 0.70, H = 20 mm). It clear that the hydrogen ratio had an obvious impact on pressure rise. As the hydrogen ratio decreased, the pressures at both ends of flame-retardant system were gradually decreased. As the flame resistance failed, the upper pressure was obviously higher than that of lower end, and the pressure difference between the two ends was decreased as the hydrogen ratio decreased. Especially the flame was quenched inside the flame-retardant system and the pressures at the two ends were almost the same as Φ = 40%. At this moment, the upper pressure was decreased from 0.44 MPa to 0.13 MPa, and the lower pressure was decreased from 0.32 MPa to 0.14 MPa. The decrease extent reached 70.5% and 56.3%, respectively.

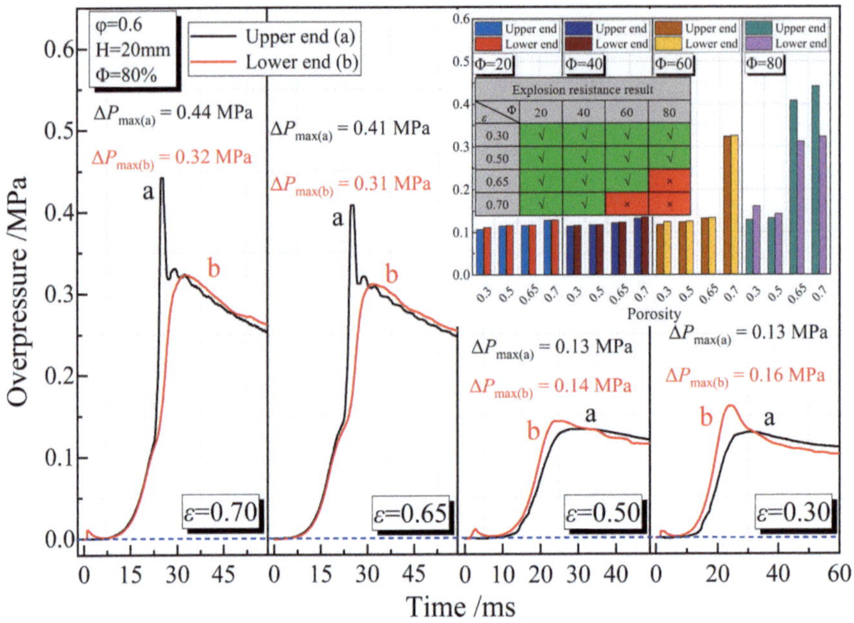

Fig. 9.8 Effect of porosity on the pressure history (φ = 0.6, Φ = 80%, H = 20 mm)

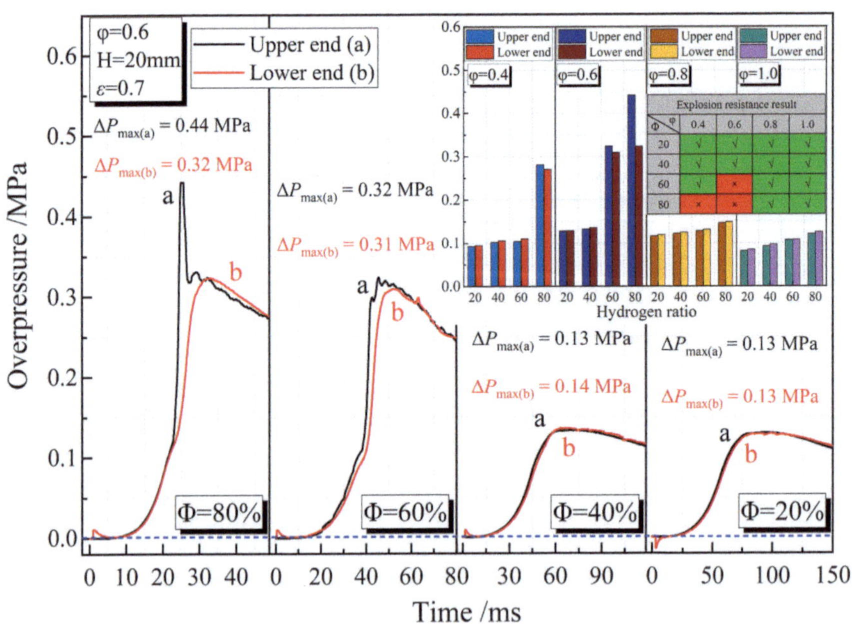

Fig. 9.9 Effect of hydrogen ratio on the pressure history (φ = 0.6, ε = 0.70, H = 20 mm)

As the hydrogen ratio continued to decrease, the change in pressure peak was not significant. It was because that the methane explosion reaction dominated as the hydrogen content dropped to a certain value, which further affected the velocity of flame propagation and intensity of shock wave entering the front-end of flame-retardant system. Besides, the pressure was the strongest and the failure probability was increased under 0.6 equivalent ratio.

9.2.4 Pressure Rising Rate

Effect of Flame-Retardant Element Thickness

The effect of flame-retardant element thickness on $(dP/dt)_{max}$ was consistent with the pressure peak ($\varphi = 0.6$, $\Phi = 80\%$), as shown in Fig. 9.10. This indicates that the increase the of element thickness could effectively reduce the explosion reaction rate and improve the flame resistance efficiency. Compared with H = 20 mm, $(dP/dt)_{max}$ at the upper and lower ends were decreased respectively by 89.44% and 60.31% as the flame resistance was successful (H = 60 mm, $\varepsilon = 0.65$). Meanwhile, $(dP/dt)_{max}$ at the upper and lower ends under the failure condition was much higher than that of the success condition, and $(dP/dt)_{max}$ of upper end was evidently higher than that of lower end. As the flame-retardant element thickness increased, the drop extent of $(dP/dt)_{max}$ at the upper end was more evidently. Especially the pressure drop became more obvious, but the change of pressure difference was not significant as the flame resistance was successful. For the lower end, $(dP/dt)_{max}$ was continuously decreased as the flame resistance failed ($\varepsilon = 0.70$). However, as the flame resistance was successful, $(dP/dt)_{max}$ was slightly enlarged and was greater than the upper end with the increase of element thickness. It was because that the resistance effect of flame-retardant system on the shock wave and flame were enhanced as the element thickness increased, resulting in a higher rate of heat accumulation than heat dissipation at the lower end. Besides, the probability of flame resistance failure was evidently enhanced as the increase of porosity and the decrease of element thickness.

Effect of Porosity

Figure 9.11 is the effect of porosity on $(dP/dt)_{max}$ at the upper and lower ends under different hydrogen ratios ($\varphi = 0.6$, H = 20 mm). It clear that $(dP/dt)_{max}$ of lower end was slightly decreased and $(dP/dt)_{max}$ of upper end was slightly enhanced under the flame resistance success with the increase of porosity, and $(dP/dt)_{max}$ of lower end was also slightly higher than that of upper end. As the hydrogen ratio decreased, $(dP/dt)_{max}$ at the upper and lower ends showed an obvious decreasing trend. Especially the decrease in $(dP/dt)_{max}$ of two ends was more significant under the flame resistance success. As $\Phi = 80\%$, $(dP/dt)_{max}$ at the upper and lower ends under 0.50 porosity (flame resistance success) were decreased respectively by 88.16% and

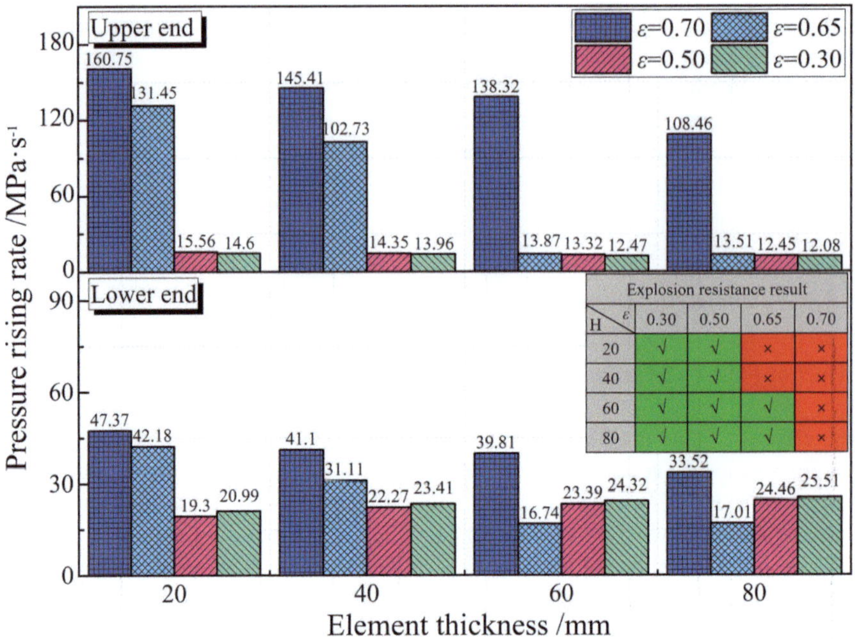

Fig. 9.10 Effect of flame-retardant element thickness on $(dP/dt)_{max}$ ($\varphi = 0.6$, $\Phi = 80\%$)

54.24% compared to 0.65 porosity. And $(dP/dt)_{max}$ at the upper and lower ends under 0.65 porosity (flame resistance failure) were decreased respectively by 18.23% and 10.96% compared to 0.70 porosity. It clear that the decrease degree under the failure condition was evidently reduced compared to the flame resistance success. Besides, the probability of flame resistance success was obviously increased as the hydrogen ratio and porosity decreased. It was because that the decrease of hydrogen ratio could reduce the intensity of shock wave, and the decrease in porosity improved the reflection effect of flame-retardant system front-end on the shock wave and the weakening degree of narrow channel, thereby improving the flame resistance efficiency [2].

Effect of Hydrogen Ratio

Figure 9.12 presents the effect of hydrogen ratio on $(dP/dt)_{max}$ under different equivalence ratios ($\varepsilon = 0.70$, $H = 20$ mm). Under same hydrogen ratio, $(dP/dt)_{max}$ presented a trend of first increasing and then decreasing with the increase of equivalence ratio. As $\varphi = 0.6$, the explosion reaction was the strongest and $(dP/dt)_{max}$ was the highest. Under the same equivalence ratio, $(dP/dt)_{max}$ also showed a decreasing trend as the hydrogen ratio decreased. Compared with the failure condition, $(dP/dt)_{max}$ of two ends was evidently reduced as the flame resistance was successful. Especially for 0.6 equivalence ratio, the flame under the condition of 40% hydrogen ratio occurred

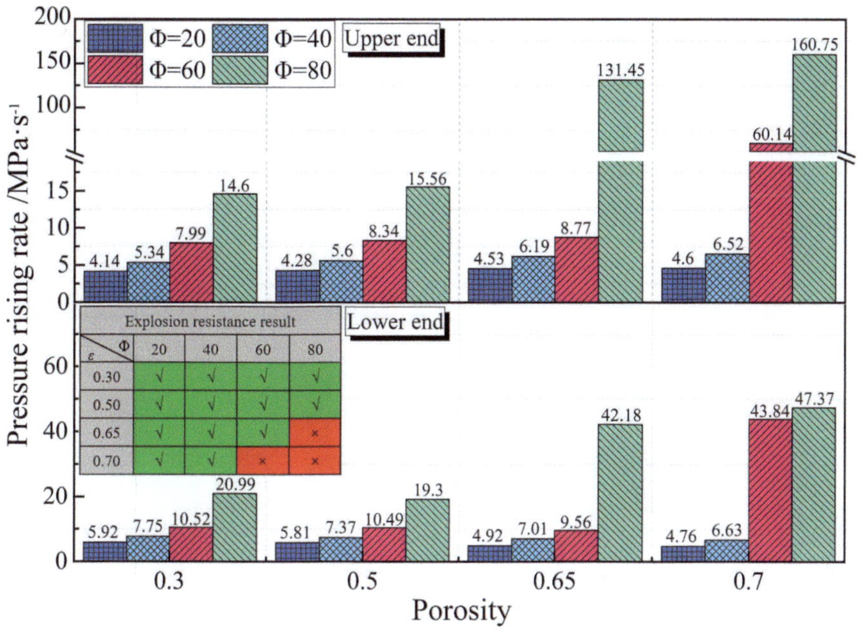

ε \ Φ	20	40	60	80
0.30	√	√	√	√
0.50	√	√	√	√
0.65	√	√	√	×
0.70	√	√	×	×

Fig. 9.11 Effect of porosity on $(dP/dt)_{max}$ (φ = 0.6, H = 20 mm)

quenching compared to 60% hydrogen ratio. And $(dP/dt)_{max}$ of upper and lower ends were decreased respectively by 89.61% and 84.88%. Under 20% hydrogen ratio, $(dP/dt)_{max}$ of two ends under 1.0 equivalence ratio were decreased respectively by 74.35% and 74.16% compared with 0.6 equivalence ratio. Under 80% hydrogen ratio, $(dP/dt)_{max}$ of two ends under 1.0 equivalence ratio were decreased respectively by 95.17% and 82.33% compared with 0.6 equivalence ratio. It clear that the hydrogen ratio had a greatly impact on the pressure rising rate of corrugated flame-retardant system two ends. As the hydrogen ratio increased, the probability of flame resistance failure was increased, and the equivalent ratio also presented a significant impact.

9.2.5 Analysis of Flame Temperature

Figure 9.13 illustrates the effect of flame-retardant element thickness on the temperature history (φ = 0.6, ε = 0.65, Φ = 80%). As the flame-retardant element thickness increased, the temperature history was greatly decreased. As the flame resistance failed (H = 20 mm and 40 mm), the temperature history at the upper end was obviously higher than that of lower end. As H = 60 mm, the flame occurred quenching. At this moment, the flame temperature of lower end was higher than that of upper end. Compared with H = 20 mm, the flame temperature peak (T_{max}) at the upper and lower ends were decreased by 71.97% and 5.56%, respectively. It clear that the

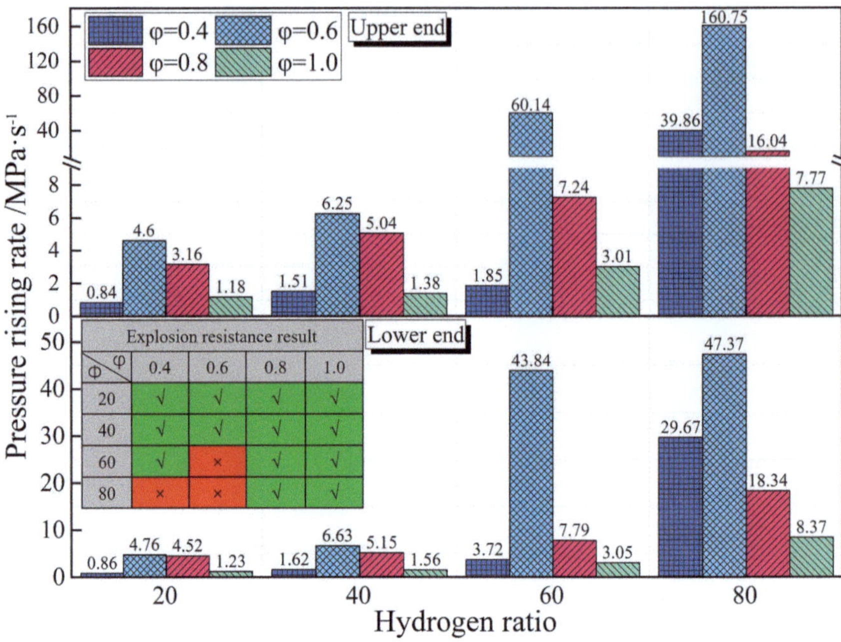

Fig. 9.12 Effect of hydrogen ratio on $(dP/dt)_{max}$ (ε = 0.70, H = 20 mm)

temperature drop at the upper end became more significant under the flame resistance success. And this also indicates that the increase of flame-retardant element thickness could improve its heat absorption efficiency on the flame, resulting in that the upper flame temperature did not reach the ignition point of premixed gas. Meanwhile, the flame temperature was decreased under the resistance failure as the porosity decreased (ε = 0.65 and 0.70), and the upper temperature history was higher with a more significant temperature drop (see in Fig. 9.14). As the porosity decreased (ε = 0.50), the upper temperature history was evidently decreased, and its decrease degree reached 70.19%. Nevertheless, the temperature history presented an upward trend as the porosity continued to decrease. It was because that the flame resistance system could effectively prevent the heat transfer, leading to an increase in the heat accumulation at the lower end and a slight increase in the lower temperature. And the heat accumulation effect was slightly increased as the porosity decreased. Figure 9.15 shows the effect of hydrogen ratio on temperature history (φ = 0.6, ε = 0.70, H = 20 mm). As the hydrogen ratio decreased, the temperature histories of two ends were evidently decreased. Compared to the failure, the drop extent of upper temperature was more obvious as the flame resistance was successful. As the flame resistance failed, the upper temperature history was much higher than that of lower end. However, the upper temperature history was smaller than that of lower end as the flame resistance was successful. And the effect of hydrogen ratio on the temperature drop was more significant compared with the element thickness and

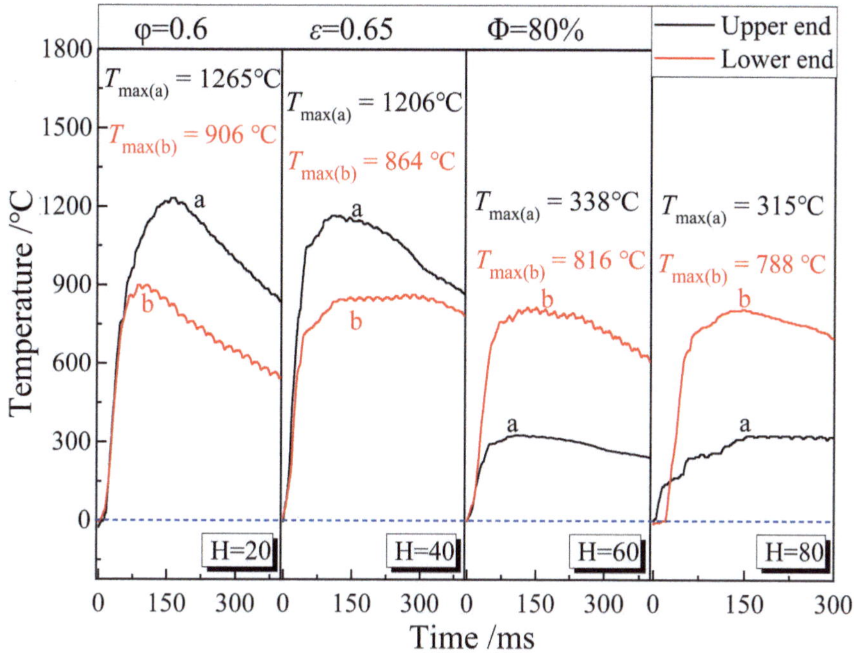

Fig. 9.13 Effect of flame-retardant element thickness on the temperature history (φ = 0.6, ε = 0.65, Φ = 80%)

porosity. Especially the flame occurred quenching as Φ = 40%. Compared with Φ = 80%, the upper temperature history was decreased by 88.93%. It was because that the flame propagation velocity entering the flame-retardant system could be altered due to the change of hydrogen ratio, thereby affecting its heat absorption rate and weakening extent inside the narrow channel. Combined with Table 9.2, the same variation pattern was also observed with the changes in the thickness and porosity of flame-retardant element and hydrogen ratio.

9.2.6 Mechanism Analysis

The quenching process of hydrogen doped methane explosion flame by the corrugated flame-retardant system could be attributed to the combined of physical and chemical effects, and the flame resistance mechanism was shown in Fig. 9.16. In terms of physical prevention, the high-speed shock wave first acted on the front-end of flame-retardant system than the high-temperature flame and partial shock wave occurred the backward propagation, weakening the intensity of shock wave and the velocity of flame propagation entering the narrow channel. Meanwhile, the discrete and compressive shock wave continuously collided with the wall surface under the

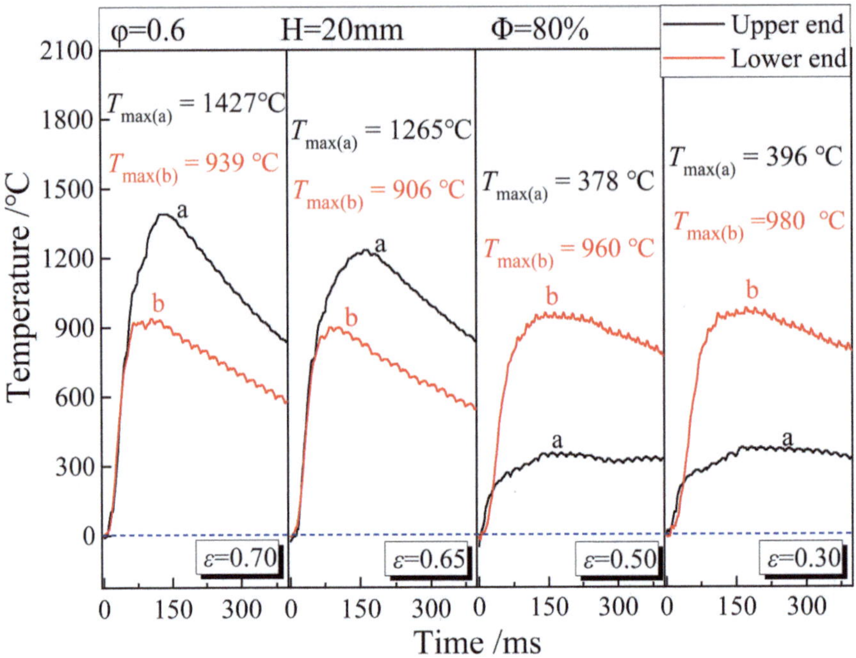

Fig. 9.14 Effect of porosity on the temperature history ($\varphi = 0.6$, $H = 20$ mm, $\Phi = 80\%$)

action of narrow channel and appeared the great attenuation. Subsequently, the high-temperature flame continuously transferred the heat to the corrugated wall surface after the discrete flame entered the narrow channel, causing the continuous decrease of flame temperature. As the front temperature dropped below the critical value [Ignition energy (H_2/CH_4) = 0.02 and 0.28 mJ], the flame would be extinguished. In addition, the accumulation of pressure would also prevent the flame propagation, thereby promoting flame quenching due to the constraint of narrow channel. In terms of chemical prevention, apart from the heat absorption by corrugated wall during the propagation of high-temperature flame, the chain reaction would also take place in the obvious change. Main reaction processes of hydrogen ($H_2 + O_2 = HO_2 + H$, $H + O_2 = O + OH$, $OH + H_2 = H_2O + H$) and methane [$O + CH_4 = CH_3 + OH$, $H + CH_3(+M) = CH_4(+M)$, $H + CH_4 = CH_3 + H_2$, $OH + CH_4 = CH_3 + H_2O$] explosions would be interrupted, leading to the great reductions in the number and energy of free radical that triggered the chain reaction [3]. As the chain reaction rate was less than the destruction rate, the flame would also occur quenching. The thickness and porosity of flame-retardant element and hydrogen ratio could change the flame resistance effect by affected the intensities of shock wave and flame entering the narrow channel, as well as the physical and chemical processes during the propagation [4].

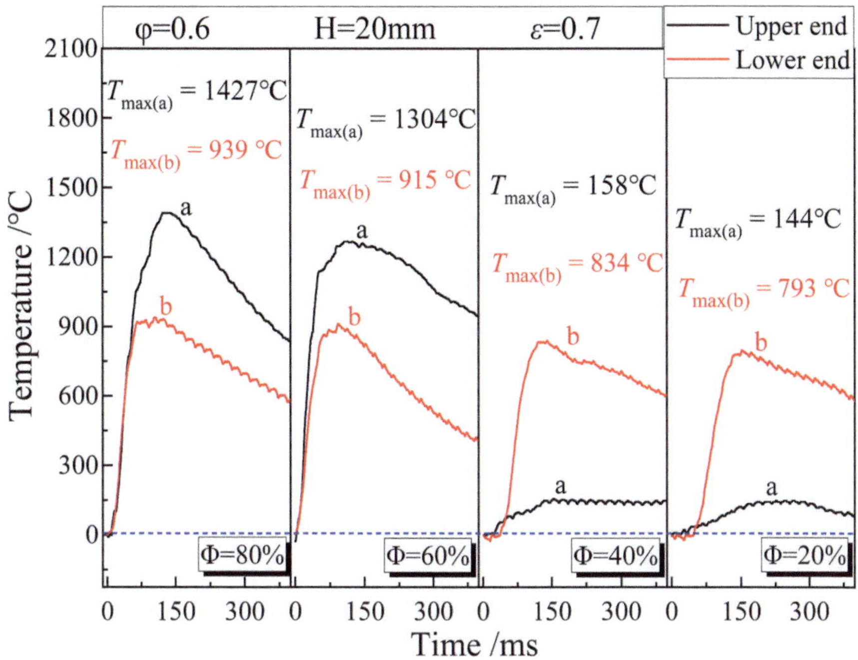

Fig. 9.15 Effect of hydrogen ratio on the temperature history ($\varphi = 0.6$, $\varepsilon = 0.70$, H = 20 mm)

9.3 Conclusions

The experiment main aimed to research the flame resistance characteristics and influence laws of corrugated flame-retardant system on hydrogen doped methane explosion. By visualizing experiment, the influence laws and mechanisms of structural parameter and hydrogen ratio on flame propagation processes and explosion parameters at two ends were clarified. The influence laws of flame propagation and shock wave attenuation inside the narrow channel was explored and the flame resistance mechanism was revealed. Conclusion are as follows:

First, there was a clear correspondence between the flame propagations and explosion parameters at the front and rear ends of the corrugated flame-retardant system. The flame propagation appeared an obvious oscillation phenomenon due to the obstruction of flame resistance system. Meanwhile, the flame propagation at the front-end could be greatly affected by the flame-retardant system, thereby affecting its propagation velocity inside the narrow channel. As the flame resistance failed, the flame at the rear-end would appear an obvious acceleration process. The structural parameter of flame-retardant system and hydrogen ratio could affect the flame resistance effect by changing the flame propagation velocity entering the narrow channel and the heat absorption rate, further weakening its intensity during the internal propagation.

Table 9.2 T_{max} under different flame resistance conditions

Explosion parameter	Temperature /°C							
Variable	Below				Above			
	φ=0.4	φ=0.6	φ=0.8	φ=1.0	φ=0.4	φ=0.6	φ=0.8	φ=1.0
Φ=20	707	792	750	642	111	144	134	93
Φ=40	741	834	771	677	128	158	140	105
Φ=60	771	915	828	755	144	1304	156	138
Φ=80	891	990	855	821	1206	1427	221	177
	Φ=20	Φ=40	Φ=60	Φ=80	Φ=20	Φ=40	Φ=60	Φ=80
ε=0.20	864	903	947	980	198	275	360	396
ε=0.25	834	885	924	960	177	254	351	378
ε=0.45	816	861	888	906	155	237	303	1265
ε=0.60	792	834	915	939	144	158	1304	1427
	ε=0.20	ε=0.25	ε=0.45	ε=0.60	ε=0.20	ε=0.25	ε=0.45	ε=0.60
H=20	939	906	960	980	396	378	1265	1427
H=40	909	864	870	891	371	365	1206	1338
H=60	879	816	828	837	359	350	338	1278
H=80	816	788	795	822	341	330	315	1175

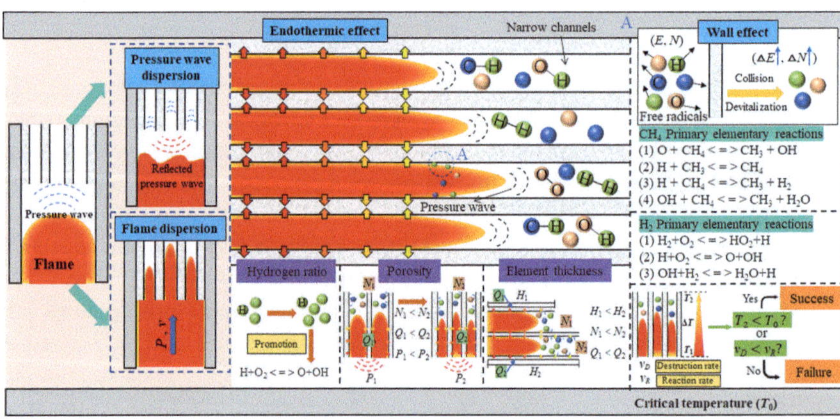

Fig. 9.16 Schematic diagram of flame resistance mechanism

Second, the corrugated flame-retardant system could effectively weaken the intensities of shock waves and reduce the explosion pressures at the front and rear ends. As the flame-retardant thickness element increased, the weakening extents of flame and shock wave by the narrow channel were increased, thereby improving the probability of flame resistance success. The decrease in porosity would increase the number of narrow channels per unit area and the dispersion degree of shock wave, leading to an enhancement in the weakening effect of shock wave during propagation inside the narrow channel and a decrease in the pressure. As the porosity and hydrogen ratio increased, the probability of flame resistance failure continued to increase.

Third, the flame temperature was greatly decreased as the flame-retardant element thickness increased. Under the failure condition, the upper flame temperature was higher than that of lower end. As the flame was quenched, the lower flame temperature was higher than that of upper end. As the porosity decreased, the flame temperature was decreased under the flame resistance failure, and the upper temperature history was higher with a significant temperature drop. As the porosity continued to decrease, the flame resistance was successful, and the temperature history of upper end was lower than that of upper end. Meanwhile, the lower flame temperature showed a slight upward trend. The increase in hydrogen ratio was not conducive to flame quenching. As the hydrogen ratio decreased, the flame temperatures at two ends were all evidently decreased. Compared to the flame resistance failure, the temperature drop of upper end flame was more significant under the success condition. The thickness and porosity of flame-retardant element and hydrogen ratio could change the flame resistance effect by affected the intensities of shock wave and flame entering the narrow channel, as well as the physical and chemical processes during the propagation.

To achieve effective resistance of hydrogen doped methane explosion flame for practical engineering application, three suggestions were proposed based on the experiment conclusion. (1) The design for structural parameter of corrugated flame-retardant system selected in the practical application should ensure the effective flame resistance, otherwise it would result in an enhancement effect and cause the greater damage. Considering the pressure drop during normal operation, its structure parameter should be optimized as much as possible. (2) For the flame resistance of high actively gas explosion, the flame-retardant system should be made of materials with high thermal conductivity. (3) To meet the practical demands of different working conditions, the quantification relationship between structural effect and explosion resistance parameters, as well as the reignition mechanism and the critical explosion resistance model should be accurately established.

References

1. Cao XY, Zhou X, Huang R, Wang Z, Wang ZR. Research on the resistance characteristics of corrugated flame-retardant system for hydrogen/methane explosion. Int J Hydrogen Energy. 2024;88:228–41.
2. Cao XY, Zhou JY, Zhou X, Wang Z, Wang ZR, Sheng YX. Experimental research on the synergy effect of resistance/inhibition on the syngas explosion. Fuel. 2024;363:130995.
3. Bao L, Wang P, Dang WY, Kuang C, Yu AF. Experimental study on detonation flame penetrating through flame arrester. J Loss Prev Process Ind. 2021;72:104529.

4. Wang LQ, Ma HH, Shen ZW. The quenching of propane deflagrations by crimped ribbon flame arrestors. J Loss Prev Process Ind. 2016;43:567–74.

Chapter 10
Effect of Obstacle Parameters on Explosion Resistance Performance of Hydrogen Crimped-Ribbon Flame Arrester

10.1 Experimental Apparatus and Procedures

Figure 10.1 illustrates a schematic diagram of experimental apparatus for studying the explosion resistance performance of hydrogen crimped-ribbon flame arrester under the presence of obstacles. The main body included the explosion pipe [including an unprotected pipe ($L_1 = 2$ m) and a protected pipe ($L_2 = 2$ m)], a flame arrester (including flame arrester element and expansion chamber structure), an obstacle group, a gas distribution system, an ignition system, a flame acquisition system, a pressure acquisition system, a temperature acquisition system and a program control and data acquisition system [1].

The explosion pipe was made of cast iron (DN65) with a design pressure of 15 MPa. Two ends of pipe were sealed by Q345R flanges. The material of flame arrester was made of 304 stainless steel. The flame arrester element was selected with the thickness (H) was 100 mm, the ripple height (λ) was 0.2 mm and the porosity (α) was 0.35, as shown in Fig. 10.2. To clarify the influence of obstacle parameters on the explosion resistance performance, a detachable obstacle group was used to change different working conditions, as shown in Fig. 10.3. The blockage rate of obstacle was calculated based on the formula ($BR = 1 - r^2/R^2$), and their distance was changed by adjusting the location of adjacent obstacles (the distance between adjacent obstacles was equal).

The premixed gas was prepared by a RCS2000-B high-precision gas distribution instrument with 1% accuracy. The ignition system consisted of a high-temperature resistant electrode and an adjustable high-energy igniter (KTD-A), with an output ignition energy range from 1 to 20 J. To measure the flame propagation velocity at the front and rear ends of flame arrester, four and one CKG100 flame detectors were installed on the unprotected pipe and the protected pipe, respectively. Two and one CY400 high-frequency pressure transducers (accuracy: 0.25% FS, range: 0–5 MPa) were installed at the front and rear ends of flame arrester to measure the pressure. Two S-type high-frequency thermocouple were installed at the front and rear ends

Z. Wang and X. Cao, *Flame Resistance Characteristics and Its Influence Mechanism*,
https://doi.org/10.1007/978-981-95-6119-3_10

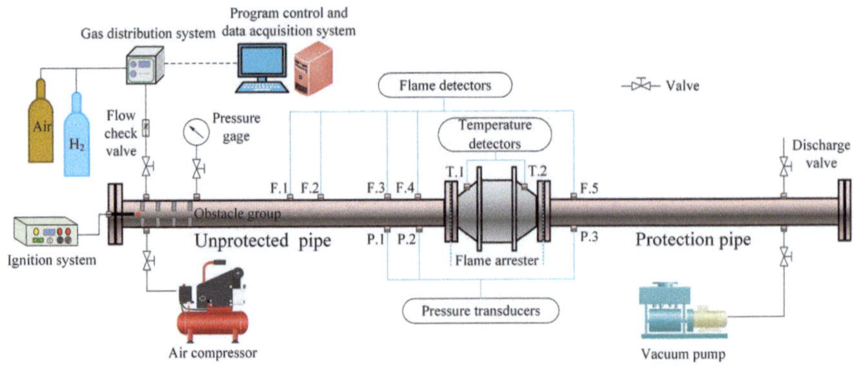

Fig. 10.1 Schematic diagram of experimental apparatus

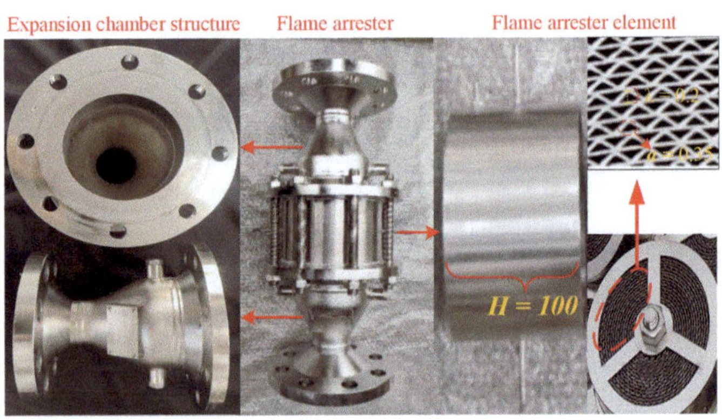

Fig. 10.2 Structure diagram of flame arrester

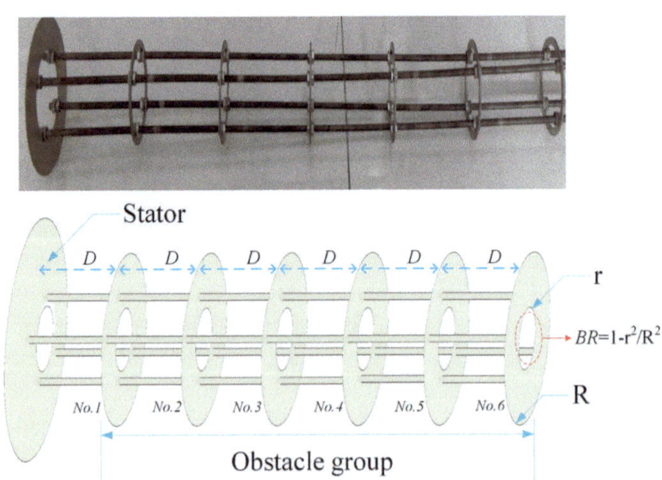

Fig. 10.3 A group diagram of obstacles

Table 10.1 Installation distance between detectors

Location 1	Location 2	Distance/mm
F.1 Flame detector	F.2 Flame detector	200
F.2 Flame detector	F.3 Flame detector	500
F.3 Flame detector	F.4 Flame detector	200
F.4 Flame detector	Left end flange of flame arrester	150
F.5 Flame detector	Right end flange of flame arrester	150
P.1 Pressure transducer	P.2 Pressure transducer	200
P.2 Pressure transducer	Left end flange of flame arrester	150
P.3 Pressure transducer	Right end flange of flame arrester	150
T.1 Temperature detector	Left end flange of flame arrester	200
T.1 Temperature detector	T.2 Temperature detector	230
T.2 Temperature detector	Right end flange of flame arrester	200

of flame arrester to measure the temperature. The specific installation location is shown in Table 10.1. Meanwhile, the acquisition system (TST6300) was controlled by program language to realize the output of program signal and data acquisition.

The premixed gas with 20% hydrogen concentration (Volume fraction) was selected. And the ignition energy was 9 J, and the initial ignition pressure was atmospheric pressure. Under the case without obstacle, the flame velocity, explosion pressure and flame temperature entering the front end of flame arrester were expressed by V_0, P_0 and T_0, respectively. As obstacles existed, the explosion parameter entering the flame arrester was expressed by V_f, P_f and T_f. And the explosion parameter at the rear end of flame arrester was expressed by V_r, P_r and T_r. The flame resistance result was expressed by R_s (S and F represent the flame resistance success and failure, respectively). Each working condition was repeated at least five times to obtain accurate and reliable experimental results. According to the results analysis, the maximum deviations of flame propagation velocity, explosion pressure and flame temperature were 8.6%, 6.9% and 2.2%, respectively, and the error bars have been added to the corresponding figures.

10.2 Results and Discussions

10.2.1 Analysis of Explosion Resistance Process

Figure 10.4 shows the voltage signals monitored by the flame detectors under flame resistance success and failure conditions, respectively (D = 120 mm; BR = 0.6; N = 2 and 8). It can be seen that the voltage signal was sharply increased as the flame passed through the flame detector (F.1–F.4). However, no obvious voltage signal was monitored by the flame detector of rear end (F.5) as the flame resistance was successful. This indicates that the flame was quenched inside the flame arrester and the premixed gas of rear end could not be ignited. The flame detector of rear end (F.5) monitored an obvious voltage signal as the flame resistance failed. This shows that the flame could pass through the flame arrester and ignite the premixed gas of rear end. The flame propagation velocity could be calculated ($V = \Delta X/\Delta t$) by the distance between adjacent flame detectors and the rise moment of voltage signal. The flame propagation velocity of front end all showed a trend of first increasing and then decreasing under above two flame resistance effects. It was because that the flame was accelerated under the action of obstacles and slowed down due to the inhibition of flame arrester. However, the flame propagation velocity under flame resistance failure was higher compared with the flame resistance success. The presence of flame arrester could significantly reduce the flame propagation velocity. In particular, the flame propagation velocity obviously decreased from 626.7 to 239.1 m/s after it passed through the flame arrester. Here, the maximum flame propagation velocity (V_{34}) near the front end of flame arrester was considered as the flame velocity (V_f) entering the flame arrester, and its value did not reach the detonation value of hydrogen (V_{CJ} = 1710 m/s) [2]. Therefore, this paper mainly focused on the explosion resistance performance of crimped-ribbon flame arrester on hydrogen deflagration.

Figure 10.5a, b present the changes of explosion pressure histories at the front and rear ends of flame arrester under flame resistance success and failure conditions, respectively ($D = 120$ mm; $BR = 0.6$; $N = 2$ and 8). It can be seen from Fig. 10.5a,

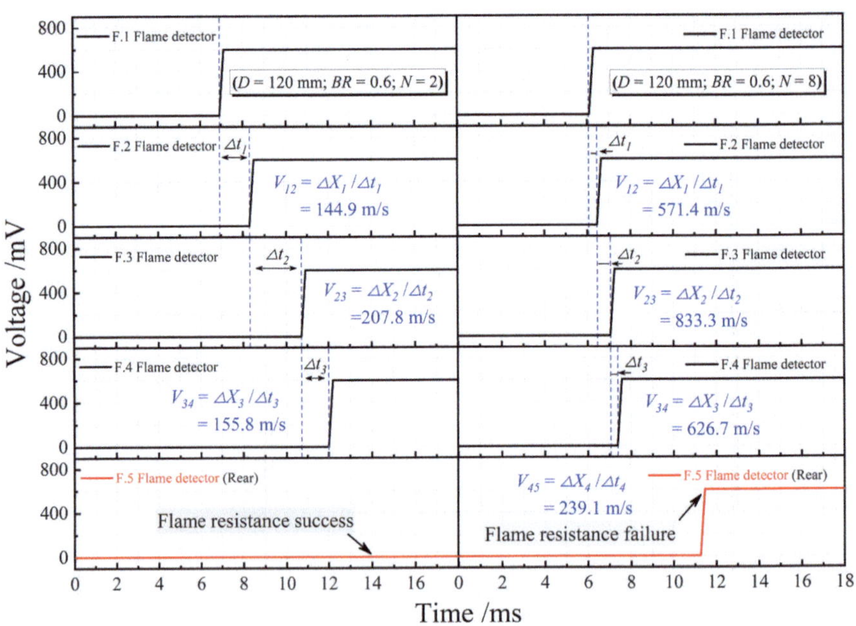

Fig. 10.4 The voltage signal monitored by the flame detectors under flame resistance success and failure conditions ($D = 120$ mm; $BR = 0.6$; $N = 2$ and 8)

the pressure rise at the front end of flame arrester could be divided into four stages when the flame resistance was successful ($D = 120$ mm; $BR = 0.6$; $N = 2$). In the first stage, the premixed gas was ignited and the flame accelerated along the pipe. However, the shock wave pressure signal had not yet reached the pressure monitoring point, resulting in no rise in the pressure history. Meanwhile, the pressure transducer (P.2) was delayed in detecting the rise moment of pressure history compared with the pressure transducer (P.1). In the second stage, the premixed gas at the front end of flame was continuously compressed by the shock wave due to the presence of flame arrester, resulting in an obvious increase of explosion pressure history. As the shock wave reached to the monitoring point, the pressure history was obviously increased (third stage). In particular, the increase of pressure near the flame arrester was more significant. Meanwhile, the combined action of the flame arrester, energy accumulation and reflected wave promoted the pressure history to rise again. In the fourth stage, the pressure showed a sharp decrease trend after the explosion flame continued to enter the flame arrester. The pressure history of rear end still showed an increase trend and appeared a peak value ($P_r = 0.07$ MPa) although the premixed gas of rear end was not ignited. This indicates that the crimped-ribbon flame arrester could not only effectively inhibit the flame propagation, but also significantly reduce the explosion intensity.

Figure 10.5b illustrates the explosion pressure histories at the front and rear ends of flame arrester under flame resistance failure condition ($D = 120$ mm; $BR = 0.6$; $N = 8$). The pressure history of front end also showed four rise stages, but there was an obvious difference. In the first stage, the flame propagation velocity was obviously increased with the increase of obstacle number, which led to an earlier rise moment in the pressure history. Meanwhile, the pressure transducer near the ignition source also showed a trend of early rise. The pressure history was sharply increased under the combined action of self acceleration and obstacle. And the pressure rise rate in the second stage was also obviously increased. After that, the pressure history was significantly increased (third stage) with the continued acceleration of flame and shock wave. Compared with the flame resistance success, the explosion pressure was obviously increased during this stage. Besides, the combined action of the flame arrester, energy accumulation and reflected wave promoted the pressure history to rise again. In the fourth stage, the pressure history of front end was sharply decreased as the explosion flame passed through the flame arrester. However, the pressure history of rear end showed an obvious increase trend and presented two peaks as the premixed gas of rear end was ignited. Compared with the flame resistance success, the pressure of rear end was obviously increased ($P_r = 0.16$ MPa). But it was obviously lower than the pressure value of front end. This was attributed to the large amount of heat and kinetic energy consumed when the flame propagated through the flame arrester. Meanwhile, the pressure transducer (P.2) was located in the area entering the flame arrester, so its monitored value was considered as the explosion pressure (P_f) entering the flame arrester.

Figure 10.6 shows the flame temperature histories at the front and rear ends of flame arrester under flame resistance success and failure conditions, respectively ($D = 120$ mm; $BR = 0.6$; $N = 2$ and 8). The temperature history of front end showed

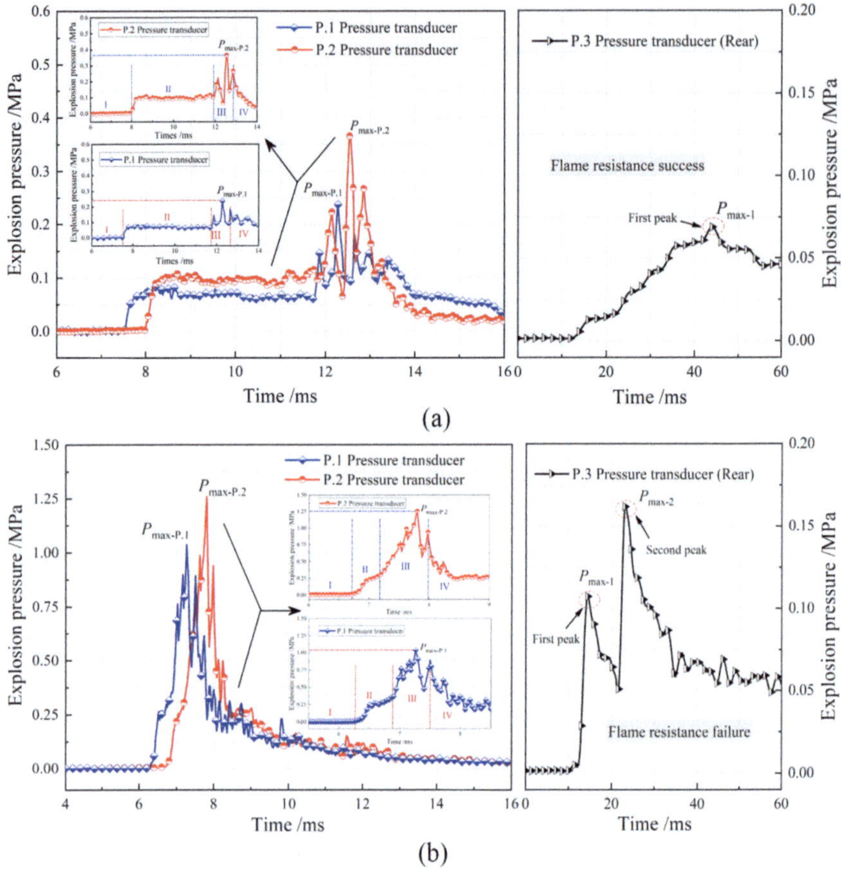

Fig. 10.5 **a** The changes of explosion pressure histories at the front and rear ends of flame arrester under flame resistance success condition (D = 120 mm; BR = 0.6; N = 2). **b** The changes of explosion pressure histories at the front and rear ends of flame arrester under flame resistance failure condition (D = 120 mm; BR = 0.6; N = 8)

an obvious increase trend due to the acceleration of flame propagation inside the pipe. The highest temperature monitored by the temperature detector (T.1) reached 1363 °C when the flame resistance was successful. However, the temperature history of rear end did not show an obvious increase trend due to the flame being quenched inside the flame arrester. Meanwhile, the flame temperature of rear end (T_r = 53 °C) did not reach the ignition temperature of premixed gas, resulting in the premixed gas of rear end not being ignited. Compared with the flame resistance success, the temperature of front end was obviously increased when the flame resistance failed (T_f = 1759 °C).

Besides, the premixed gas of rear end was ignited as the flame passed through the flame arrester, further resulting in an increase in the flame temperature of rear end (T_r = 872 °C). The flame temperature of front and rear ends all showed a decrease

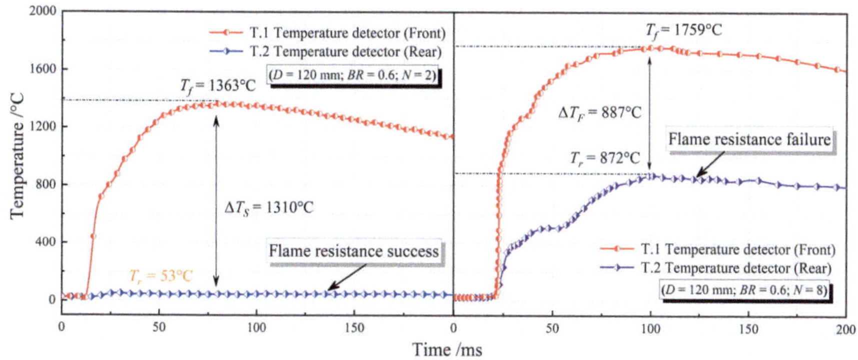

Fig. 10.6 The changes of flame temperature histories at the front and rear ends of flame arrester under flame resistance failure and failure conditions ($D = 120$ mm; $BR = 0.6$; $N = 2$ and 8)

trend under above two flame resistance effects due to the presence of flame arrester. However, the temperature difference of flame resistance success was greater ($\Delta T_S > \Delta T_F$) compared with the flame resistance failure. This indicates that the presence of flame arrester could significantly weaken the explosion intensity and absorb the heat generated by the flame. In particular, the effect of heat absorption was more significant when the flame resistance was successful.

10.2.2 *Analysis of Obstacle Influencing Factors*

Effect of Obstacle Location

Figure 10.7 presents the flame propagation velocity at the front and rear ends of flame arrester under different obstacle locations ($N = 1$; $BR = 0.7$). It can be seen that the flame propagation velocity showed a trend of first increasing and then decreasing with the increase of obstacle location. The flame propagation velocity reached the maximum as L was 650 mm. In particular, the flame propagation velocity of front end was 5.2 times than that under the case without obstacle ($V_f = 5.2V_0$). The flame propagation inside the pipe was accelerated under the action of obstacle ($V_f > V_0$). However, the flame acceleration was less affected by the obstacle as the obstacle was close to the ignition source ($L < 325$ mm). The flame quenching inside the flame arrester under this condition, and no obvious voltage signal was monitored by the flame detector of rear end. As the obstacle was far away from the ignition source (325 mm $< L <$ 845 mm), the flame propagation velocity was obviously increased and it could pass through the flame arrester under the combined action of self-acceleration and obstacle [3]. Thereafter, the flame acceleration effect began to decrease as the obstacle location continued to increase ($L > 845$ mm). This further led to a decrease in the flame velocity entering the flame arrester. Based on the flame

Fig. 10.7 The flame propagation velocity under different obstacle locations ($N = 1$; $BR = 0.7$)

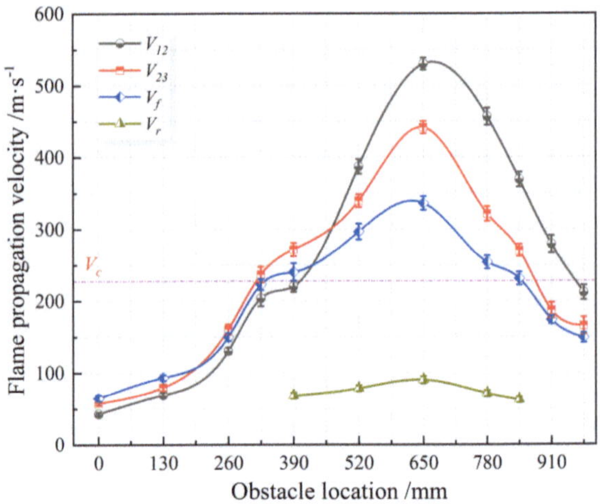

resistance results and the corresponding flame propagation velocity, the critical flame resistance velocity was determined ($V_c = 227.8$ m/s). Meanwhile, the flame velocity showed a decrease trend due to the blocking effect of flame arrester ($V_r < V_f$).

Figure 10.8 illustrates the explosion pressure near the front and rear ends of flame arrester under different obstacle locations ($N = 1$; $BR = 0.7$). The explosion pressure of front end showed a trend of first increasing and then decreasing with the increase of obstacle location. The pressure of front end reached the maximum as L was 650 mm ($P_f = 4.0P_0$). However, there was an obvious difference in the explosion pressure of rear end [4]. As the obstacle was close to the ignition source ($L < 325$ mm), the explosion pressure was slightly increased under the action of obstacle. The explosion shock wave intensity inside the flame arrester was obviously weakened under this condition, further resulting in the pressure history of rear end presented one peak. As the obstacle was far away from the ignition source (325 mm < L < 845 mm), the explosion pressure was obviously increased under the combined action of flame self-acceleration and obstacle. This led to the flame resistance failure and the pressure history of rear end appeared two peaks. Besides, the shock wave inside the narrow channel was greatly weakened during propagation, which led to a lower peak in the pressure history I when the flame resistance failed. The peak of pressure history II showed the same trend as the explosion pressure of front end during this range. After that, the promotion effect of obstacle on the shock wave was weakened as the obstacle location continued to increase ($L > 845$ mm). This led to a decrease in the explosion pressure entering the flame arrester and the flame resistance was successful. Meanwhile, the critical flame resistance pressure was determined ($P_c =$ 0.45 MPa) based on the flame resistance results and the corresponding explosion pressure.

Figure 10.9 shows the flame temperature at the front and rear ends of flame arrester and the corresponding flame resistance results under different obstacle locations (N = 1; $BR = 0.7$). It can be seen that the flame temperature showed a trend of first

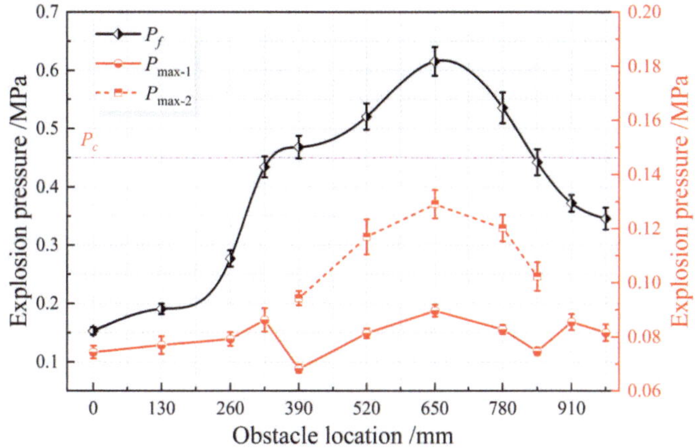

Fig. 10.8 The explosion pressure under different obstacle locations ($N = 1$; $BR = 0.7$)

increasing and then decreasing with the increase of obstacle location. As L was 650 mm, the flame temperature at the front and rear ends all reached the maximum and the T_f was greater than T_r. However, the flame flow field was less disturbed by obstacle as the obstacle was close to the ignition source ($L < 325$ mm). The flame temperature entering the flame arrester was low under this condition, which resulted in no obvious increase in the temperature of rear end. This indicated that the flame resistance was successful. As the obstacle was far away from the ignition source (325 mm $< L <$ 845 mm), the flame temperature was obviously increased under the combined action of self-acceleration and obstacle. In particular, the flame could pass through the flame arrester and ignite the premixed gas of rear end, further resulting in the flame resistance failure and an obvious increase in the temperature of rear end. The disturbance of flame flow field by the obstacle began to decrease as the obstacle location continued to increase ($L >$ 845 mm), which led to the decrease of flame temperature entering the flame arrester. In particular, the temperature history of rear end did not show an obvious increase trend.

According to the flame resistance results, the flame resistance failure began to occur as L was 325 mm, and the probability of flame resistance success showed a trend of first decreasing and then increasing with the increase of obstacle location. As 520 mm $\leq L \leq$ 780 mm, the flame resistance all failed. However, the phenomenon of flame resistance success was presented again as L was 845 mm. It can be seen that the probability of flame resistance success was related to the obstacle location and was greatly affected by it. Besides, based on explosion parameters and flame resistance results, the quantitative prediction models of V_f, P_f and T_f under different obstacle locations were proposed, as shown in formulas (10.1–10.3).

$$V_f = 48.61414 + (174{,}806.42809/(498.00896 \times \sqrt{\pi/2})) \times \exp\left(-2 \times ((L - 602.29066)/498.00896)^2\right)$$

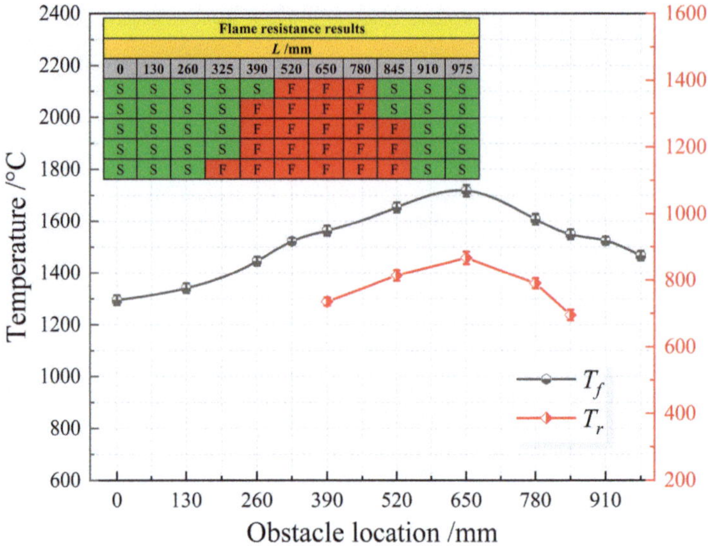

Fig. 10.9 Temperature and flame resistance results under different obstacle locations (N = 1; BR = 0.7)

$$R^2 = 0.98678 \tag{10.1}$$

$$\begin{aligned} P_f &= 0.12109 + (315.63988/(522.11371 \times \sqrt{\pi/2})) \\ &\quad \times \exp\left(-2 \times ((L - 622.48957)/522.11371)^2\right) \\ R^2 &= 0.97581 \end{aligned} \tag{10.2}$$

$$\begin{aligned} T_f &= 1247.02629 + (321{,}198.21616/(581.08771 \times \sqrt{\pi/2})) \\ &\quad \times \exp(-2 \times ((L - 622.74745)/581.08771)^2) \\ R^2 &= 0.98433 \end{aligned} \tag{10.3}$$

Effect of Obstacle Distance

Figure 10.10a–c present the influence of obstacle distance on the explosion parameters and the corresponding flame resistance results under different obstacle numbers (BR = 0.7; N = 4, 6 and 8). The explosion parameters (V_f, P_f and T_f) showed a trend of first increasing and then slowly decreasing with the increase of obstacle distance [5]. As N was 4 and 6, the maximum explosion parameters entering the flame arrester were obtained at the distance of 100 mm. As N was 8, the maximum

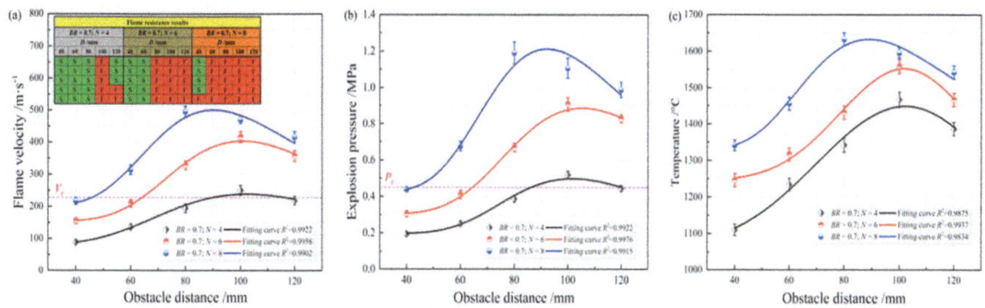

Fig. 10.10 Effects of obstacle distance on V_f, P_f, T_f and R_s under different obstacle numbers (*BR* = 0.7; *N* = 4, 6 and 8)

explosion parameters were obtained at the distance of 80 mm. The minimum explosion parameters were all obtained at the distance of 40 mm under three obstacle numbers. Meanwhile, the explosion parameter was continuously increased with the increase of obstacle number. Combined with the flame velocity, explosion pressure and flame temperature, the effect of obstacle distance on the flame resistance result was obviously different when the obstacle number was changed. As *N* was 4, the flame resistance failure occurred at $D \geq 100$ mm. And the probability of flame resistance success at the distance of 100 mm was the lowest. However, the probability of flame success resistance was continuously decreased with the increase of obstacle number. As *N* was 6, the flame resistance failure occurred at $D \geq 80$ mm. As *N* was 8, the flame resistance failure occurred at $D \geq 40$ mm. In summary, the obstacle distance could significantly affect the explosion parameters entering the flame arrester, which further affected the probability of flame resistance success. And the larger explosion parameter was more likely to appear when the obstacle distance was larger, which indicated that the flame resistance failure mostly occurred at the larger distance.

Effect of Obstacle Number

Figure 10.11a–c illustrate the influence of obstacle number on the explosion parameters and the corresponding flame resistance results under different blockage rates (*D* = 120 mm; *BR* = 0.5, 0.6 and 0.7). The explosion parameters (V_f, P_f and T_f) showed an increase trend with the increase of obstacle number, and the rising degree was continuously increased. In particular, the explosion parameters reached the maximum under the action of obstacle number was 8.

The influence of blockage rate on the explosion parameter was not obvious as $N < 4$. However, the blockage rate had an obvious effect on the explosion parameter as $N \geq 4$. The above two obstacle parameters affected the probability of flame resistance success by affecting the explosion parameter. The flame resistance results were successful under three blockage rates as $N < 4$. The flame resistance results were failed under three blocking rates as $N > 4$. The probability of flame success resistance was increased with the increase of blockage rate as *N* was 4. This indicates

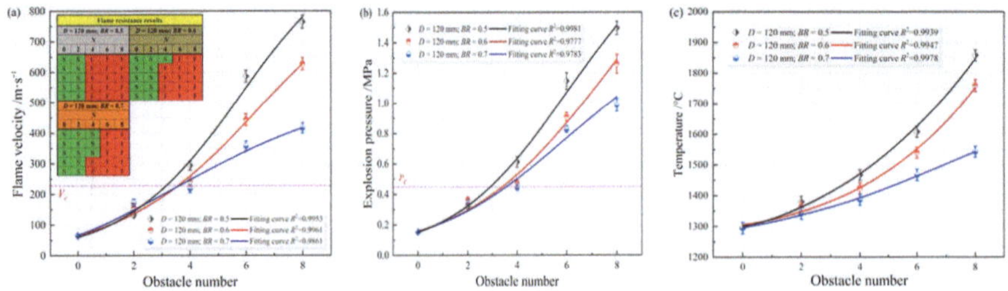

Fig. 10.11 Effects of obstacle number on V_f, P_f, T_f and R_s under different obstacle blockage rates ($D = 120$ mm; $BR = 0.5$, 0.6 and 0.7)

that the obstacle number could significantly affect the explosion parameter entering the flame arrester, and then affect the flame resistance result. In particular, the flame resistance failure mostly occurred when the obstacle number was larger.

Effect of Blockage Rate

Figure 10.12a–c show the influence of blockage rate on the explosion parameters and the corresponding flame resistance results under three obstacle numbers ($D =$ 120 mm; $N = 4$, 6 and 8). The explosion parameters (V_f, P_f and T_f) near the front end of flame arrester showed a trend of first increasing and then decreasing with the increase of blockage rate. In particular, the explosion parameters reached the maximum under the action of obstacles as N was 8 and BR was 0.5 ($V_f =$ $11.8V_0$, $P_f = 9.8P_0$ and $T_f = 1.4T_0$). And the probability of flame success resistance was the lowest. Besides, the explosion parameter was continuously increased with the increase of obstacle number. Combined with the explosion parameters, the probability of flame success resistance showed a trend of first decreasing and then increasing with the increase of blockage rate as N was 4. However, the flame resistance all failed under five blockage rates as the obstacle number continued to increase ($N = 6$ and 8). This indicates that the blockage rate could significantly affect the explosion parameters entering the flame arrester, and then affect the probability of flame resistance success. In particular, the flame resistance failure was more likely to occur when the blockage rate was larger.

10.2.3 Analysis of Quenching Mechanism

The endothermic effect and wall effect were the two main mechanisms of hydrogen explosion flame quenching inside the narrow channel of crimped-ribbon flame arrester. Figure 10.13 presents a schematic diagram of hydrogen explosion flame quenching process inside the narrow channel of flame arrester. It can be seen that the explosion flame accelerated and reached the expansion chamber under the action

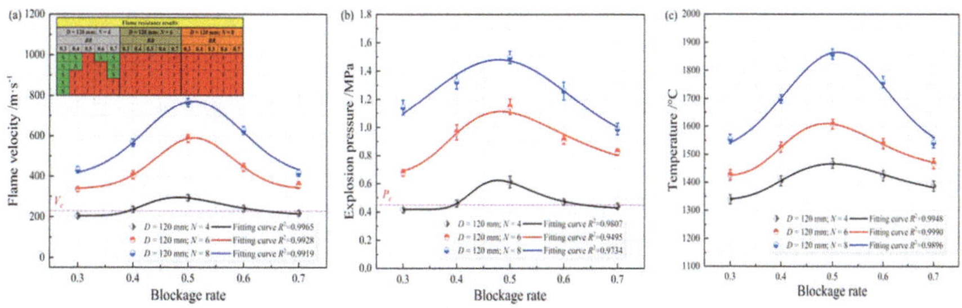

Fig. 10.12 Effects of blockage rate on V_f, P_f, T_f and R_s under different obstacle numbers ($D =$ 120 mm; $N = 4$, 6 and 8)

of obstacles. Firstly, part of the shock wave collided with the flame arrester element and generated the reflected wave at different angles, causing the loss of part energy and the reduction of explosion intensity. Meanwhile, the reflected wave acted on the incoming flame, causing the disturbance of flow field inside the expansion chamber. Then, a vortex motion was formed inside the expansion chamber during this process, which slowed down the flame propagation velocity [6]. Besides, part of the shock wave could enter and reach the rear end of flame arrester through the flame arrester element. The shock wave formed the scattered at different angles inside the narrow channel, causing the strong weakening of explosion shock wave and an obvious decrease in the pressure of rear end. The continuous flame was dispersed into discrete flame structure and entered the narrow channels due to the blocking effect of flame arrester element. The flame continuously lost heat during propagation due to the endothermic effect of wall. This further led to a decrease in the flame temperature and below the critical value, and the flame quenching occurred. After flame quenching, a small amount of heat was still transferred to the rear end, further resulting in a slight increase in the temperature of rear end.

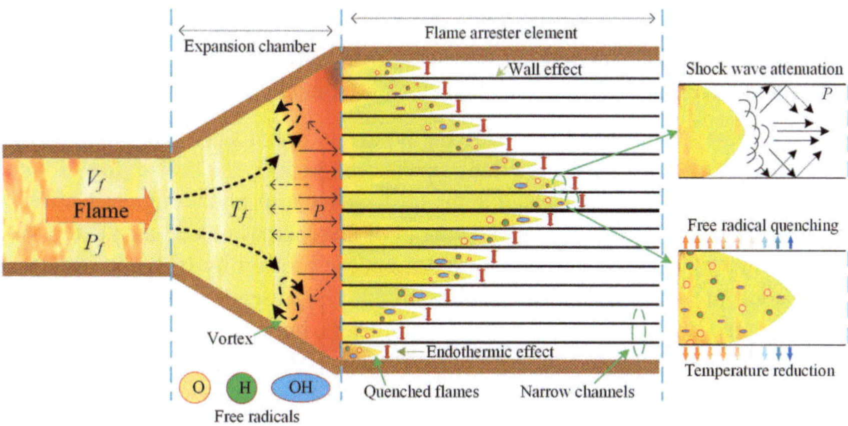

Fig. 10.13 Quenching mechanism of hydrogen explosion flame inside the narrow channel

Besides, the wall effect of flame arrester element on the hydrogen explosion flame was also an important factor for flame quenching. There were many chain reaction processes during hydrogen explosion. The main reaction processes were as follows ($H + O_2 = O + OH$, $O + H_2 = H + OH$, $OH + H_2 = H + H_2O$, $2OH = O + H_2O$). However, the free radicals collided with the wall during flame propagation. The capacity and quantity of free radicals were continuously decreased during the collision and contact. As the number of producing free radicals was less than the number of consuming free radicals, the number of free radicals could not support the chain reaction, further resulting in flame quenching. Under the combined action of endothermic and wall effect, the flame front temperature and the number of chain reaction free radicals were reduced to a critical value, thereby achieving effective flame resistance. The presence of obstacles could greatly affect the flame velocity, explosion pressure and flame temperature entering the narrow channel of flame arrester, which affected the explosion resistance effect. Meanwhile, the influence degree of obstacle number on the explosion resistance performance was more significant. Increasing the obstacle number could obviously increase the explosion intensity, which was favorable for the flame propagation through the flame arrester. Under the synergistic action of endothermic and wall effect, there was a critical quenching value for the flame arrester. The flame resistance failed when the explosion parameter entering the flame arrester was greater than the critical quenching value.

10.3 Conclusions

The explosion resistance performance of hydrogen crimped-ribbon flame arrester under different obstacle location, distance, number and blockage rate were studied, and the influence laws of obstacle parameters on V_f, P_f and T_f entering the flame arrester were investigated. Meanwhile, the flame resistance result of crimped-ribbon flame arrester on hydrogen explosion under the action of obstacle was revealed, and the quenching mechanism of flame arrester and the influence mechanism of obstacle parameters were clarified. The main conclusions could be summarized as follows:

(1) The presence of obstacles inside the pipe could obviously change the explosion parameters entering the flame arrester. Moreover, the probability of flame resistance success was related to the obstacle location and was greatly affected by it. The V_f, P_f and T_f showed a trend of first increasing and then decreasing with the increase of obstacle location, but the probability of flame resistance success presented an opposite change trend. The presence of obstacles not only promoted the increase of explosion parameters, but also led to the flame resistance failure. When the flame resistance failed, a flame rise signal was monitored by the flame detector of rear end and two peaks appeared in the pressure history of rear end. It explained that the flame velocity and explosion pressure entering the flame arrester were the key factors affecting the flame resistance result.

Based on the flame resistance results and the corresponding flame velocity and explosion pressure, the critical flame resistance velocity and explosion pressure were determined.

(2) The explosion parameter showed a trend of first increasing and then decreasing with the increase of obstacle distance. And the larger explosion parameter was more likely to appear when the obstacle distance was larger, which indicated that the flame resistance failure mostly occurred at the larger distance. Meanwhile, the probability of flame resistance success was decreased with the increase of obstacle number. This was attributed to the fact that the explosion parameter was increased with the increase of obstacle number, and the rising degree was continuously increased. The obstacle number could significantly affect the explosion parameter entering the flame arrester, and then affect the flame resistance result. In particular, the flame resistance failure mostly occurred when the obstacle number was larger. Besides, the blockage rate could significantly affect the explosion parameter entering the flame arrester, thereby affecting the probability of flame resistance success. In particular, the probability of flame resistance success showed a trend of first decreasing and then increasing with the increase of blockage rate as N was 4. However, the flame resistance all failed under five blockage rates as the obstacle number continued to increase ($N = 6$ and 8). The explosion intensity entering the flame arrester was obviously increased under the action of obstacles, which further affected the flame resistance effect of flame arrester on the hydrogen explosion. The flame resistance failed when the explosion parameter entering the flame arrester was greater than the critical quenching value.

The findings of this study are of great significance to improve and supplement the explosion resistance performance of flame arrester under the condition of obstacles and to enhance the safety and reliability of the pipe system. In order to effectively inhibit the explosion of hydrogen-air premixed gas inside the pipe and improve the flame resistance effect of hydrogen crimped-ribbon flame arrester, three suggestions are proposed based on the above work: (1) The interior of pipe where the flame arrester was installed should be free of obstacles and smooth. Otherwise, the number, distance, and blockage rate of obstacles should be minimized. (2) In order to effectively reduce the explosion parameters entering the flame arrester, the expansion chamber structure should be optimized to greatly weaken the explosion intensity. (3) In order to improve the flame resistance performance, the flame resistance material with high thermal conductivity should be selected preferentially under the condition of ensuring normal gas transportation.

References

1. Wang JZ, Cao XY, Wang ZR, Lin CD, Shen YA. Effect of obstacle parameters on explosion resistance performance of hydrogen crimped-ribbon flame arrester. Int J Hydrogen Energy. 2023;48:34440–53.

2. Malik K, Zbikowski M, Bak D, Lesiak P, Teodorczyk A. Numerical and experimental investigation of H_2-air and H_2O_2 detonation parameters in a 9 m long tube, introduction of a new detonation model. Int J Hydrogen Energy. 2019;44(17):8743–50.
3. Han SX, Yu MG, Yang XF, Wang XY. Effects of obstacle position and hydrogen volume fraction on premixed syngas air flame acceleration. Int J Hydrogen Energy. 2020;45(53):29518–32.
4. Lv XS, Zheng L, Zhang YG, Yu MG, Su Y. Combined effects of obstacle position and equivalence ratio on overpressure of premixed hydrogen-air explosion. Int J Hydrogen Energy. 2016;41(39):17740–9.
5. Na'inna AM, Phylaktou HN, Andrews GE. Prediction of distance to maximum intensity of turbulence generated by grid plate obstacles in explosion-induced flows. J Loss Prev Process Ind. 2021;69:104318.
6. Chu JW, Liu Y, Lv X, Li QP, Dong HS, Song YC, et al. Experimental investigation on blockage predictions in gas pipelines using the pressure pulse wave method. Energy. 2021;230:120897.

Chapter 11
Research on Quenching Performance and Multi-factor Influence Law of Hydrogen Crimped-Ribbon Flame Arrester Using Response Surface Methodology

11.1 Experimental Apparatus and Procedures

The schematic diagram of the experimental apparatus was shown in Fig. 11.1. According to the ISO 16852 standard, the experimental apparatus was built, including an unprotected pipe ($L_1 = 2$ m), the flame arrester (flame arrester element and expansion chamber structure), a protected pipe ($L_2 = 2$ m), a high-voltage ignition system, a gas distribution system, a pressure acquisition system, a flame detection system, a data acquisition system, and a synchronous controller [1].

The unprotected and protected pipes were made of cast iron (DN65) with a design pressure of 15 MPa. The flame arrester was made of 304 stainless steel. Figure 11.2 showed the flame arrester with three porosities ($\alpha = 0.35, 0.5, 0.65$) and three element thicknesses ($H = 60, 80, 100$ mm). In order to effectively reduce the flame intensity acting on the flame arrester and analyze the influence of different expansion chambers, three types of expansion chamber structures (the extended type expansion chamber, the baffle type expansion chamber, and the conventional type expansion chamber) were used to connect to the flame arrester element, as shown in Fig. 11.3. The three expansion chamber structures have the same volume, but differ only in their internal structure. A ring-type baffle with an inner diameter of 70 mm was installed inside the baffle-type expansion chamber structure, and the distances of the baffle from the flame arrester element and flange end face were 50 mm and 215 mm respectively. The extended type expansion chamber structure was equipped with 155 mm pipe, and a 12 mm diameter round hole evenly distributed around it.

The high-voltage ignition system consisted of a high-temperature resistant ignition electrode and a KTD-A adjustable high energy igniter with an output ignition energy range from 1 to 20 J. The RCS2000-B high-precision gas distribution equipment was used for premixed gas preparation, and the gas distribution accuracy was 1%. The CY400 high-frequency pressure transducers (accuracy: 0.25% FS; range: 0–1.5 MPa) were used and installed on the unprotected pipe 400 mm and 200 mm from the left end of the flame arrester, respectively. The CKG100 flame detectors were used

Z. Wang and X. Cao, *Flame Resistance Characteristics and Its Influence Mechanism*,
https://doi.org/10.1007/978-981-95-6119-3_11

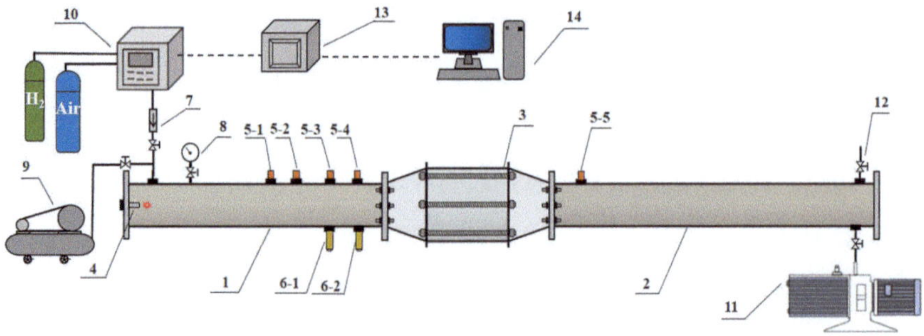

1-Unprotected pipe; 2-Protected pipe; 3-Flame arrester; 4-Ignition electrodes; 5-Flame detectors; 6-Pressure transducers; 7-Check valve; 8-Pressure gauge; 9-Air compressor; 10-Gas distribution system; 11-Vacuum pump; 12-Exhaust valve; 13-Synchronous controller; 14-Program control and acquisition system

Fig. 11.1 Schematic diagram of experimental apparatus

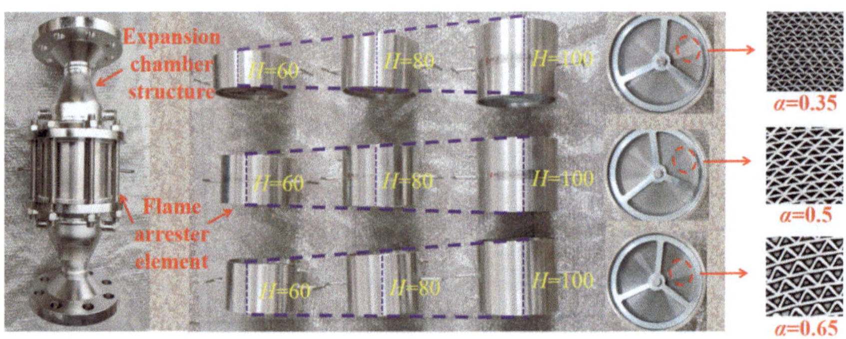

Fig. 11.2 Flame arresters with different porosities and element thicknesses

and installed in the unprotected and protected pipes respectively. The No. 1–4 flame detectors were installed on the unprotected pipe, and the distances from the entrance of the flame arrester were 1100 mm, 900 mm, 400 mm, and 200 mm respectively. The No. 5 flame detector was installed on the protected pipe 200 mm from the right end of the flame arrester to detect the success or failure of the flame quenching. The TST6300 data acquisition instrument and synchronous control system were used to realize the program control and data acquisition.

The experiments were carried out with the hydrogen-air premixed gas. The initial pressure (P_0), hydrogen concentration (C_0) and ignition energy (E_0) could be adjusted by the gas distribution system and high-voltage ignition system. The research schemes for the effects of flame arrester structure parameters and initial conditions on flame quenching performance were shown in Tables 11.1 and 11.2 respectively. Then high-voltage discharge ignition and data acquisition were controlled cooperatively by the data acquisition system and the synchronous controller. To achieve reliable results, each experiment was repeated five times at least in this research. According to the measured results, the maximum deviations

(a) Conventional type expansion chamber housing (b) Baffle type expansion chamber housing (c) Extended type expansion chamber housing

Fig. 11.3 Three types of expansion chamber structures

of the maximum explosion pressure and flame propagation velocity were 5.4% and 7.9% respectively, and the error bars have been added to the relevant figures.

In addition, based on the experimental results of single-factor effects on flame quenching performance, the effects of multi-factor interactions on the flame propagation velocity and explosion pressure entering the expansion chamber of the flame arrester were investigated using RSM. In the multi-factor interactions analysis of initial conditions, the Box-Behnken experiment design scheme (BBD) with 3-factor 3-level was used, and the experimental scheme was shown in Table 11.3.

In the multi-factor interaction analysis of structure parameters, the Central Composite experiment design scheme (CCD) with 2-factor 3-level was used [2], and the experimental scheme was shown in Table 11.4. In addition, the multi-factor prediction models of the critical flame propagation velocity and explosion pressure under different structure parameters were established respectively, and the critical flame quenching criterion of flame arrester was proposed.

Table 11.1 Research scheme for the effect of initial conditions on the flame quenching performance

No.	Influencing factors		Other initial conditions	Structure parameter values of flame arrester
A1	C_0	10%	$P_0 = 0$ MPa $E_0 = 13$ J	$\alpha = 0.35$ $H = 80$ mm
A2		20%		
A3		30%		
A4		40%		
A5		50%		
B1	P_0	– 0.05 MPa	$C_0 = 30\%$ $E_0 = 13$ J	
B2		– 0.025 MPa		
B3		0 MPa		
B4		0.025 MPa		
B5		0.05 MPa		
C1	E_0	3 J	$C_0 = 30\%$ $P_0 = 0$ MPa	
C2		8 J		
C3		13 J		
C4		18 J		

Table 11.2 Scheme for effect of flame arrester structure parameters on flame quenching performance

No	Influencing factors		Other structure parameters	Initial conditions
D1	α	0.35	$H = 100$ mm	$C_0 = 30\%$ $P_0 = 0$ MPa $E_0 = 13$ J
D2		0.5		
D3		0.65		
E1	H	60 mm	$\alpha = 0.35$	
E2		80 mm		
E3		100 mm		

Table 11.3 The BBD experiment design of initial conditions

No	C_0 (%)	P_0 (MPa)	E_0 (J)	No	C_0 (%)	P_0 (MPa)	E_0 (J)
1	20	– 0.05	13	10	30	0.05	8
2	40	– 0.05	13	11	30	– 0.05	18
3	20	0.05	13	12	30	0.05	18
4	40	0.05	13	13	30	0	13
5	20	0	8	14	30	0	13
6	40	0	8	15	30	0	13
7	20	0	18	16	30	0	13
8	40	0	18	17	30	0	13
9	30	– 0.05	8				

Table 11.4 The CCD experiment design of structure parameters

No	A	H (mm)	No	α	H (mm)
1	0.35	60	8	0.5	100
2	0.65	60	9	0.5	80
3	0.35	100	10	0.5	80
4	0.65	100	11	0.5	80
5	0.35	80	12	0.5	80
6	0.65	80	13	0.5	80
7	0.5	60			

11.2 Results and Discussions

11.2.1 Influencing Factors on Flame Quenching Performance

Initial Conditions

Figures 11.4 and 11.5 show the flame signal and explosion pressure curves measured at different positions respectively. As can be seen from Fig. 11.4, when the explosion flame passes the flame detector position, the flame detector detects the flame, which causes the electrical signal curve to rise sharply. The flame propagation velocity is calculated from the distance of adjacent flame detectors and the flame signal interval time. By comparing the flame propagation velocity in different sections, it can be seen that the flame propagation velocity at the front end of the flame arrester increases with the distance between the detectors and the flame arrester decreasing. Meanwhile, the flame propagation velocity is significantly reduced after the explosion flame enters the flame arrester compared the maximum flame propagation velocity before and after the flame arrester, which indicates that the flame arrester can effectively reduce the flame propagation velocity entering the flame arrester.

Figure 11.5 shows the explosion pressure curves measured at different positions near the front of the flame arrester. It can be seen that the process of the rising pressure curve at the front end of the flame arrester can be divided into four stages. In the I stage, the explosion pressure has no obvious change, because the flame propagation ability is weak, and the affected area is small in the initial stage of the explosion reaction. The remote pressure transducers do not feel the pressure wave, so the pressure curves in the I stage do not change. In the II stage, the explosion reaction rate increases with the explosion flame propagating along the pipe, and the pressure curve under the action of pressure wave shows a small upward change. With the accelerated propagation of the flame inside the pipe, the explosion pressure increases significantly. When the flame pass through the pressure transducers, the explosion pressure curves collected by the two transducers show a sharp rise in turn, and the maximum explosion pressure values are 0.46 MPa and 0.64 MPa respectively, which indicates that the explosion

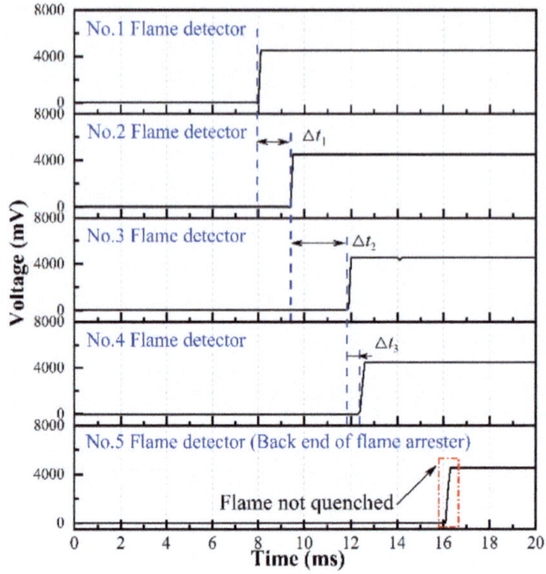

Fig. 11.4 The flame signals collected by the flame detectors

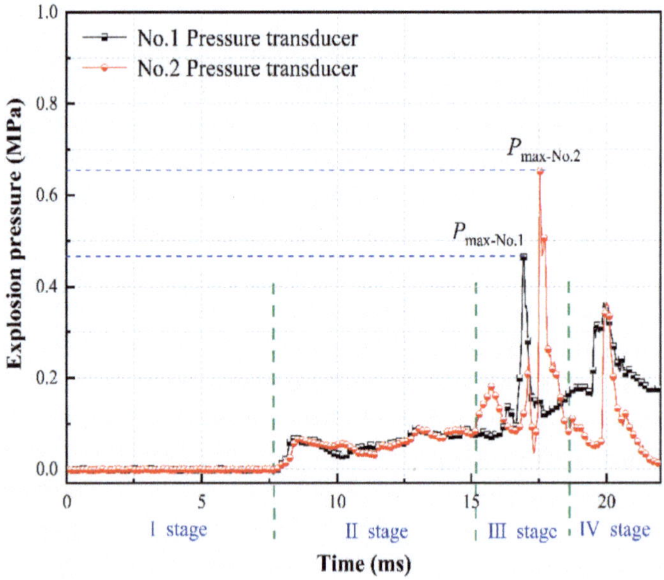

Fig. 11.5 The explosion pressure curves collected by pressure transducers

pressure curve measured by the transducer close to the flame arrester rises more obviously (III stage). When the explosion flame enters the expansion chamber and contacts with the flame arrester element, part of the flame and pressure wave could pass through the narrow channels. And the part of the pressure wave will propagate backward under the action of the flame arrester element, which leads to a secondary rise in the pressure curve, and the peak pressure appears at the same time (IV stage). However, because only part of the pressure wave passes through the flame arrester, the secondary pressure peak value is significantly reduced. The explosion pressure value is evaluated by acoustic theory and the maximum explosion pressure value of 0.30 MPa is calculated based on the acoustic theory [3]. However, the predicted value is less than the experimentally measured value ($P_{\text{max-No. 1}} = 0.46$ MPa) in the pipe. And the experimental value is close to the Sun's experimental value in confined space [4]. This is because the explosion flame propagates along the pipe due to the restraint of wall surface compared to the gas explosion in the open space. The flame propagation is accelerated, which leads to the increase of the explosion pressure in the pipe. By comparing Figs. 11.4 and 11.5, it can be found that the maximum flame propagation velocity and explosion pressure value do not reach the detonation value of hydrogen-air premixed gas ($V_{CJ} = 1970$ m/s, $P_{CJ} = 1.54$ MPa) [5], so the experiment is the process of quenching the hydrogen deflagration flame by the crimped-ribbon pipe flame arrester. In addition, the No. 4 flame detector and No. 2 pressure transducer are located in the area entering the expansion chamber, so the maximum flame propagation velocity and explosion pressure measured at this location are regarded as the flame propagation velocity (V_f) and explosion pressure (P_f) entering the flame arrester respectively.

Figure 11.6a–c show the V_f, P_f, and the corresponding flame quenching results under different initial pressure (P_0), hydrogen concentration (C_0), and ignition energy (E_0) respectively. The experimental scheme is shown in Table 11.1. As shown in Fig. 11.6a, the V_f and P_f show a variation trend of first increasing and then decreasing with increasing the C_0. It can be observed that when the C_0 is 30%, which is close to the chemical equivalent concentration of hydrogen explosion, and the explosion intensity is the largest. The V_f and P_f reached the maximum values of 248 m/s and 0.45 MPa respectively, and the flame quenching failure at this time. The V_f and P_f values decrease with the C_0 moving away from 30%, and the explosion flame is successfully quenched. Based on the experimental results, the critical Peclet numbers of crimped ribbon flame arrester are calculated and evaluated for different equivalence ratios. It can be found that the critical Peclet number is around 14, which is close to the Peclet number (21–28) calculated according to the MESG values in ISO 16852 [6]. This indicates that the flame arrester has better flame quenching performance for hydrogen explosion flame, which is consistent with the experimental results.

Meanwhile, it can be seen from Fig. 11.6b that both V_f and P_f increase significantly with increasing the P_0. Especially when the P_0 is 0.05 MPa, the V_f and P_f values reach 280 m/s and 1.04 MPa respectively. According to the experimental results, when the P_0 is − 0.05 MPa and − 0.025 MPa, the explosion flame is effectively quenched. And the flame arrester fails when the P_0 is positive. This is because

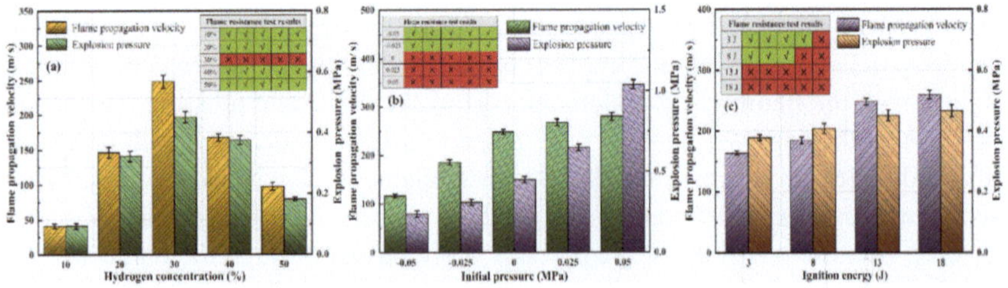

Fig. 11.6 The V_f, P_f, and flame quenching results under different initial conditions (√—no flame transmission, ×—flame transmission)

the hydrogen content per unit volume increases with increasing the P_0. The higher the chemical energy released per unit volume of hydrogen molecules when the premixed gas is ignited, and the hydrogen explosion intensity increases. Therefore, the V_f and P_f values of hydrogen explosion increase with increasing the P_0, and the probability of flame quenching failure increases. In addition, the V_f and P_f also show an increasing trend with increasing the E_0. Especially when the E_0 is 18 J, the V_f and P_f reach 259 m/s and 0.47 MPa respectively, as shown in Fig. 11.6c. According to the experimental results, the probability of successful flame quenching increases gradually when the E_0 is < 13 J, and the flame arrester fails when the E_0 is greater than 13 J. This is because the energy to excite the chain reaction increases and the number of radicals involved in the reaction increases with increasing the E_0, resulting in an increase in the explosion reaction rate and the explosion intensity, so the explosion flame is difficult to be quenched. As can be seen above, the flame propagation velocity and explosion pressure entering the flame arrester are important factors that influence the success of the flame arrester. However, the initial conditions indirectly affect the flame quenching results by directly influencing the flame propagation velocity and explosion pressure entering the flame arrester.

Structure Parameters of Flame Arrester

Figure 11.7a, b show the V_f, P_f, and the corresponding flame quenching results under different porosities (α) and element thicknesses (H) respectively. The experimental scheme is shown in Table 11.2. As shown in Fig. 11.7a, it can be observed that the V_f and P_f values increase progressively with increasing the α, this result is consistent with the experimental result of Jin [7]. The reflected pressure wave acting with the flame surface is significantly enhanced with decreasing the porosity, and the increase in flame turbulence makes the V_f and P_f values increased. Meanwhile, the probability of flame quenching success increases significantly with decreasing the α. The flame arrester fails when the α is greater than 0.5, and the flame quenching success occurs when the α is 0.35. In addition, the V_f value decreases progressively (from 273 to 211 m/s) with increasing the H value, but the P_f value increases gradually (from 0.37 to 0.65 MPa), as shown in Fig. 11.7b. This is because relatively high flow resistance

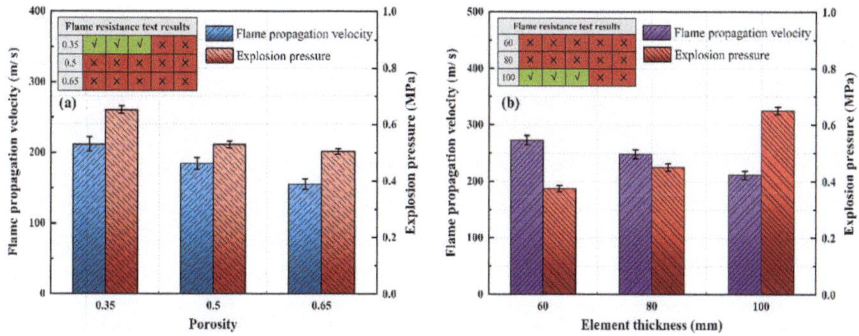

Fig. 11.7 The V_f, P_f, and flame quenching results under different structure parameters (√—no flame transmission, ×—flame transmission)

is produced by increasing the H, resulting in the higher pressure and lower flame propagation velocity in the front end of the flame arrester. The flame arrester fails when the H is < 80 mm, and the flame quenching success of flame arrester occurs when the H is 100 mm, which indicates that the increase in the H can improve the flame quenching performance. With the H increasing, although the P_f value increases gradually, the V_f value decreases, resulting in the phenomenon of successful flame quenching by flame arrester. In summary, it can be seen that the structure parameters of flame arrester (the porosity and the element thickness) are related to the flame quenching performance of flame arrester. The changes in the structure parameters can affect the flame propagation velocity and explosion pressure entering the flame arrester, and could also affect the flame propagation process in the narrow channels of flame arrester, which in turn affects the flame quenching results.

Expansion Chamber Structure

The expansion chamber indirectly affects the flame quenching results by influencing the flame propagation velocity and explosion pressure entering its interior. Figure 11.8a, b show the V_f, P_f, and the corresponding flame quenching results under different expansion chamber structures. The V_f and P_f show a variation trend of increasing and then decreasing with increasing the hydrogen concentration. Meanwhile, compared with the conventional type expansion chamber structure, the extended type and the baffle type expansion chamber have better flame quenching performance. Besides, the influence order of the three expansion chamber structures on the V_f and P_f is: the extended type expansion chamber > the baffle type expansion chamber > the conventional type expansion chamber. It can be discovered that the V_f and P_f are availably reduced by the baffle type and extended type expansion chambers.

As a result of the endothermic effect and the wall effect, the crimped-ribbon flame arrester can quench hydrogen explosion flame [8]. Figure 11.9 shows the

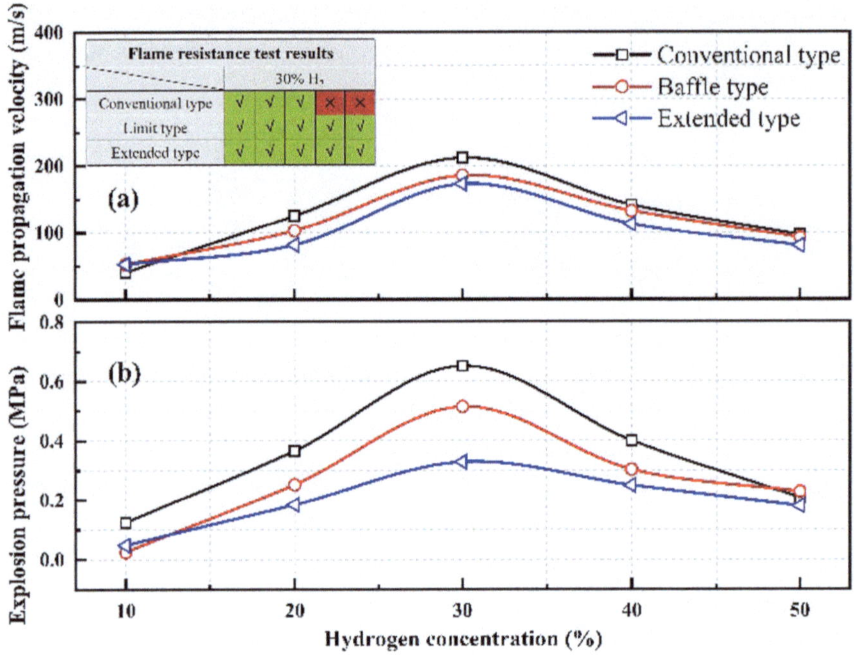

Fig. 11.8 The V_f, P_f, and flame quenching results under different expansion chamber structures (√—no flame transmission, ×—flame transmission)

flame quenching process in the expansion chamber and narrow channels of the flame arrester. When the flame enters narrow channels, it is separated into several small flames and propagates in the narrow channels. The explosion flame propagation in the narrow channel could generate heat loss, resulting in a decrease in the flame front temperature. And the flame is quenched when the temperature decreases to a critical value. Meanwhile, it is known from the theory of explosion chain reactions that a large number of free radicals ($H + O_2 = O + OH$, $O + H_2 = H + OH$, $OH + H_2 = H + H_2O$, $2OH = O + H_2O$) are generated in the hydrogen explosion elementary reaction, and free radicals can participate in and contribute to the chain reaction. However, the free radicals will be destroyed by collision with the narrow channel walls, which leads to a decrease in the number of free radicals. When the number of radical destruction is greater than the number of radicals generated that can participate in the reaction, the explosion chain reaction no longer proceeds, and the flame is quenched. Meanwhile, the initial conditions can affect the flame propagation velocity and explosion pressure entering the flame arrester, which in turn affect the heat transfer and quenching process of the flame inside the narrow channels. When the explosion flame entering the flame arrester has a large propagation velocity and explosion pressure, the propagation time of the explosion flame in the narrow channels decreases. This would weaken the endothermic effect and wall effect of the narrow channels on the explosion flame, and the flame front temperature and the free radical number of the chain reaction do not fall below the critical value of flame quenching, resulting in the continued propagation of the explosion flame. The hydrogen content per unit volume increases with increasing the hydrogen concentration and initial pressure, which further leads to an increase in the number of radicals per unit volume involved in the explosion elementary reaction inside the narrow channels. The reduction in the free radical number caused by the wall effect does completely terminate the chain reaction, resulting in the continued propagation of the explosion flame. Besides, the structure parameters can affect the flame propagation velocity and explosion pressure entering the flame arrester. The increase in the element thickness could increase the narrow channel length, and the interaction time between the explosion flame and the wall surface of the narrow channel would increase. Thus, the endothermic effect and the wall effect of the narrow channels on the explosion flame are enhanced, resulting in the successful quenching of the explosion flame. Besides, the explosion flame is divided into a smaller flame stream with decreasing the porosity. The contact area between the small flame and the narrow channel wall increases, and the heat transfer between the explosion flame and the wall is enhanced. The probability of collision and destruction between free radicals and the wall is enhanced, so that the explosion flame is quenched [9].

In addition, the pressure wave diffusion occurs when the flame enters the baffle type expansion chamber, the pressure waves on both sides are blocked by the baffle plate and then propagate in reverse, and the reflected pressure waves are attenuated after colliding with the expansion chamber structure. The pressure wave passing through the ring-type baffle diffuses again under the action of the baffle edge, which significantly reduces the intensity of the pressure wave acting on the flame arrester element. Meanwhile, due to the blocking effect of the ring-type baffle, the intensity of the reflected pressure wave propagating back to the unprotected pipe is significantly reduced. The disturbance intensity to the flame flow field in the unprotected pipe is weakened, which leads to the decrease of V_f and P_f. In addition, after the pressure wave enters the extended type expansion chamber, the pressure wave would propagate radially when it propagates to the surrounding evenly distributed small holes, and attenuates after interacting with the expansion chamber structure. Affected by the cross-sectional area of the pipe, the pressure wave propagates back to the unprotected pipe, and acts on the flame surface to make flame turbulent is significantly reduced, which in turn causes the V_f and P_f values to be reduced [10]. Therefore, the expansion chamber structure can effectively reduce the flame propagation velocity and explosion pressure entering the flame arrester, and also can increase the collision between the explosion flame and the housing, which would increase the energy loss of the explosion flame, resulting in the explosion flame being quenched more easily. In summary, it can be seen that the flame quenching performance is influenced by multiple factors such as the structure parameters, the initial conditions, and the expansion chamber structures. The three influencing factors can indirectly affect the flame quenching results by affecting the flame propagation velocity and explosion pressure entering the flame arrester, as well as the endothermic effect and wall effect of the narrow channels.

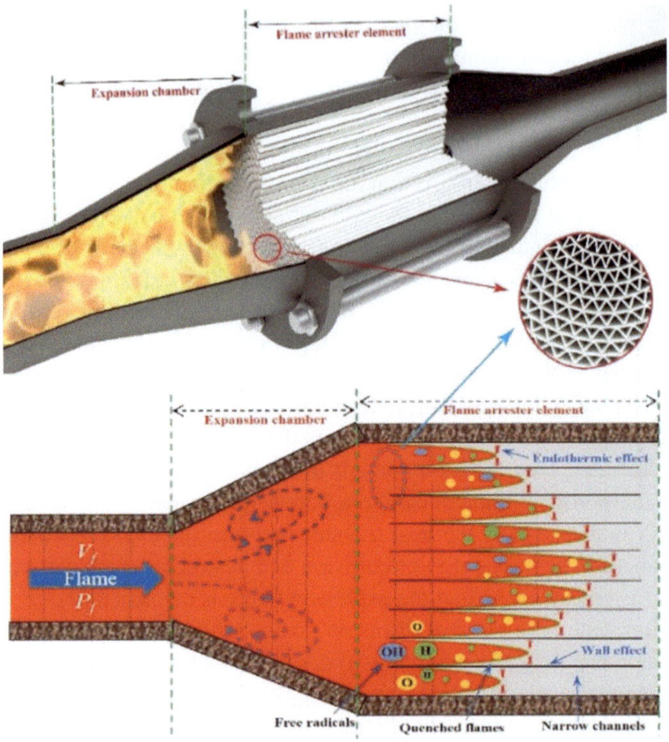

Fig. 11.9 A mechanism by which the flame arrester succeeds. The explosion flame is quenched in the expansion chamber and narrow channels of flame arrester due to endothermic effect and wall effect

11.2.2 *Interactions of Multiple Factors on Flame Quenching Performance*

Interactions of Initial Conditions

Based on single-factor influence analysis, the interaction effects of initial conditions on the flame quenching performance are studied by RSM. The Box-Behnken experiment design scheme (BBD) with 3-factor 3-level is used, as shown in Table 11.3. The response parameters are the flame propagation velocity (V_f) and the explosion pressure (P_f) entering the flame arrester, and the design parameters are the initial pressure (P_0), hydrogen concentration (C_0), and ignition energy (E_0) respectively. Meanwhile, the multi-factor prediction models of V_f and P_f under different initial conditions are obtained by RSM as follows:

$$\begin{aligned} V_f = & -331.62732 + 31.55253 * C_0 + 1417.37250 * P_0 \\ & + 11.00443 * E_0 + 13.44730 * C_0 P_0 + 0.054585 * C_0 E_0 \end{aligned}$$

$$
\begin{aligned}
&- 13.11200 * P_0E_0 - 0.517358 * C_0^2 - 15440.22000 * P_0^2 \\
&- 0.401310 * E_0^2 \\
R^2 &= 0.9984
\end{aligned} \tag{11.1}
$$

$$
\begin{aligned}
P_f =& -0.717351 + 0.070455 * C_0 + 7.57750 * P_0 + 0.006029 * E_0 \\
&- 0.028000 * C_0P_0 - 0.000190 * C_0E_0 + 0.055000 * P_0E_0 \\
&- 0.001085 * C_0^2 + 69.91000 * P_0^2 + 0.000171 * E_0^2 \\
R^2 =& 0.9989
\end{aligned} \tag{11.2}
$$

To verify the validity of the prediction models (Table 11.5), and determine the influence of the influencing factors and their interactions on the response parameters, the analysis of variance (ANOVA) is used to analyze the error of the models and the influence of each factor. Tables 11.5 and 11.6 show the ANOVA results of prediction models of the V_f and P_f respectively. It can be found that the error probability (P-value) of the two prediction models is < 0.0001, the models are "significant", and the Lack of Fit of both models is "not significant". Meanwhile, the R^2 values of the models are both greater than 0.90. In addition, as can be seen in Fig. 11.10, the predicted values of the models are in good agreement with experimental data. In summary, it is shown that the prediction models of V_f and P_f have good accuracy. The smaller the P-value the greater the influence of the influencing factor on the response parameters [11]. It can be found from ANOVA that the P-values of P_0, C_0, and E_0 are < 0.05, which indicates that all three influencing factors have a significant influence on V_f and P_f. Since the P-values of all three influencing factors are < 0.0001, and the larger the F-value, the greater the influence of the design parameter on the response parameters, so the influence extent is analyzed by comparing the F-values corresponding to the design parameters. By comparing the F-value, it can be obtained that the influence order of the three factors in both models is: $P_0 > C_0 > E_0$.

Besides, it found from ANOVA that the interaction effects of initial conditions also have a significant effect on the prediction models. Figure 11.11a shows the interaction effects of C_0 and P_0 on V_f and P_f respectively. It can be seen that when the C_0 is the equivalent value, both V_f and P_f increase significantly with increasing the P_0. When the C_0 is far from the equivalent value, both V_f and P_f show an increasing trend with increasing the P_0. Figure 11.11b shows the interaction effects of C_0 and E_0 on V_f and P_f respectively. With increasing the C_0 and E_0, both V_f and P_f show a trend of increasing first and then decreasing, and the C_0 has a greater influence on V_f and P_f than the E_0. Figure 11.11c shows the interaction effects of P_0 and E_0 on V_f and P_f respectively. V_f and P_f values increase with increasing the E_0 and P_0. Compared with the E_0, the P_0 has a more significant effect on V_f and P_f values. According to Tables 11.5 and 11.6, comparing the P-values of each two factors' interaction in the initial conditions, it is found that the influence order of interaction effects on V_f and P_f is: (C_0 and P_0) > (P_0 and E_0) > (C_0 and E_0).

Table 11.5 ANOVA for the response surface regression model of the V_f

Source	Sum of Squares	df	Mean Square	F-value	P-value	
Model	76,332.09	9	8481.34	1206.32	< 0.0001	Significant
C_0	1192	1	1192	169.54	< 0.0001	
P_0	54,472.15	1	54,472.15	7747.72	< 0.0001	
E_0	974.99	1	974.99	138.67	< 0.0001	
C_0P_0	180.83	1	180.83	25.72	0.0014	
C_0E_0	29.8	1	29.8	4.24	0.0785	
P_0E_0	42.98	1	42.98	6.11	0.0427	
C_0^2	11,269.87	1	11,269.87	1602.94	< 0.0001	
P_0^2	6273.69	1	6273.69	892.32	< 0.0001	
E_0^2	423.82	1	423.82	60.28	0.0001	
Residual	49.22	7	7.03			
Lack of fit	27.6	3	9.2	1.7	0.3033	Not significant
Pure error	21.61	4	5.4			
Cor total	76,381.3	16				

Table 11.6 ANOVA for the response surface regression model of the P_f

Source	Sum of squares	df	Mean Square	F-value	P-value	
Model	1.29	9	0.1438	2308.95	< 0.0001	Significant
C_0	0.0067	1	0.0067	108.01	< 0.0001	
P_0	1.11	1	1.11	17,831.82	< 0.0001	
E_0	0.0046	1	0.0046	73.2	< 0.0001	
C_0P_0	0.0008	1	0.0008	12.59	0.0094	
C_0E_0	0.0004	1	0.0004	5.8	0.047	
P_0E_0	0.0008	1	0.0008	12.14	0.0102	
C_0^2	0.0495	1	0.0495	795.35	< 0.0001	
P_0^2	0.1286	1	0.1286	2064.7	< 0.0001	
E_0^2	0.0001	1	0.0001	1.24	0.3031	
Residual	0.0004	7	0.0001			
Lack of fit	0.0002	3	0.0001	0.879	0.5232	Not significant
Pure error	0.0003	4	0.0001			
Cor total	1.29	16				

Interactions of Structure Parameters

Based on single-factor influence analysis, the interaction effects of structure parameters on the flame quenching performance are studied by RSM. The Central

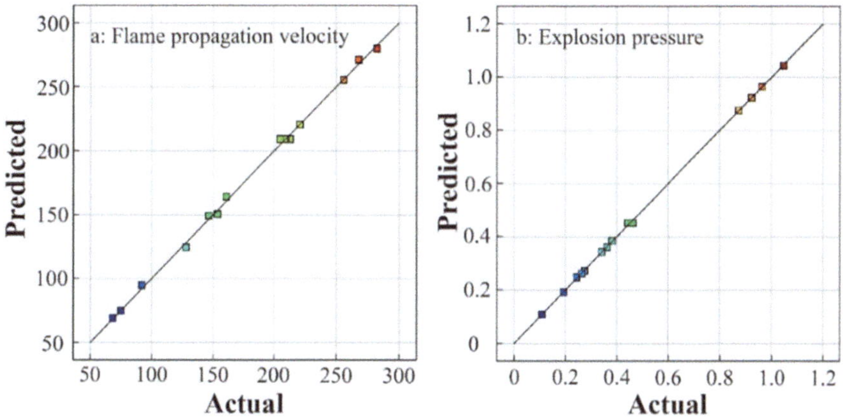

Fig. 11.10 The predicted versus actual values plot for V_f and P_f

Composite experiment design scheme (CCD) with 2-factor 3-level is used, as shown in Table 11.4. The response parameters are the flame propagation velocity (V_f) and the explosion pressure (P_f) entering the flame arrester, and the design parameters are flame arrester porosity (α) and element thickness (H). Meanwhile, the multi-factor prediction models of V_f and P_f under different structure parameters are obtained by RSM as follows:

$$\begin{aligned} V_f &= -617.06505 - 325.72632 * \alpha - 5.48886 * H \\ &\quad - 3.09970 * \alpha H + 343.41176 * \alpha^2 + 0.035203 * H^2 \\ R^2 &= 0.9902 \end{aligned} \tag{11.3}$$

$$\begin{aligned} P_f &= 0.697762 - 0.501303 * \alpha - 0.005935 * H \\ &\quad - 0.007083 * \alpha H + 0.727969 * \alpha^2 + 0.000087 * H^2 \\ R^2 &= 0.9921 \end{aligned} \tag{11.4}$$

According to the P-values corresponding to the V_f and P_f prediction models in ANOVA, and the comparison results between the predicted and experimental values in Fig. 11.12, as well as the R^2 value of the equations, it can be found that the prediction models of V_f and P_f have good accuracy with the experimental values. The P-values of structure parameters are < 0.05 according to the ANOVA results, which indicates that α and H have a significant influence on V_f and P_f. By comparing the F-values corresponding to the design parameters, it can be found that the influence of α (F-value = 377.46) on V_f model is greater than that of H (F-value = 250.20), while the influence of H (F-value = 588.64) on P_f. model is greater than that of α (F-value = 191.14).

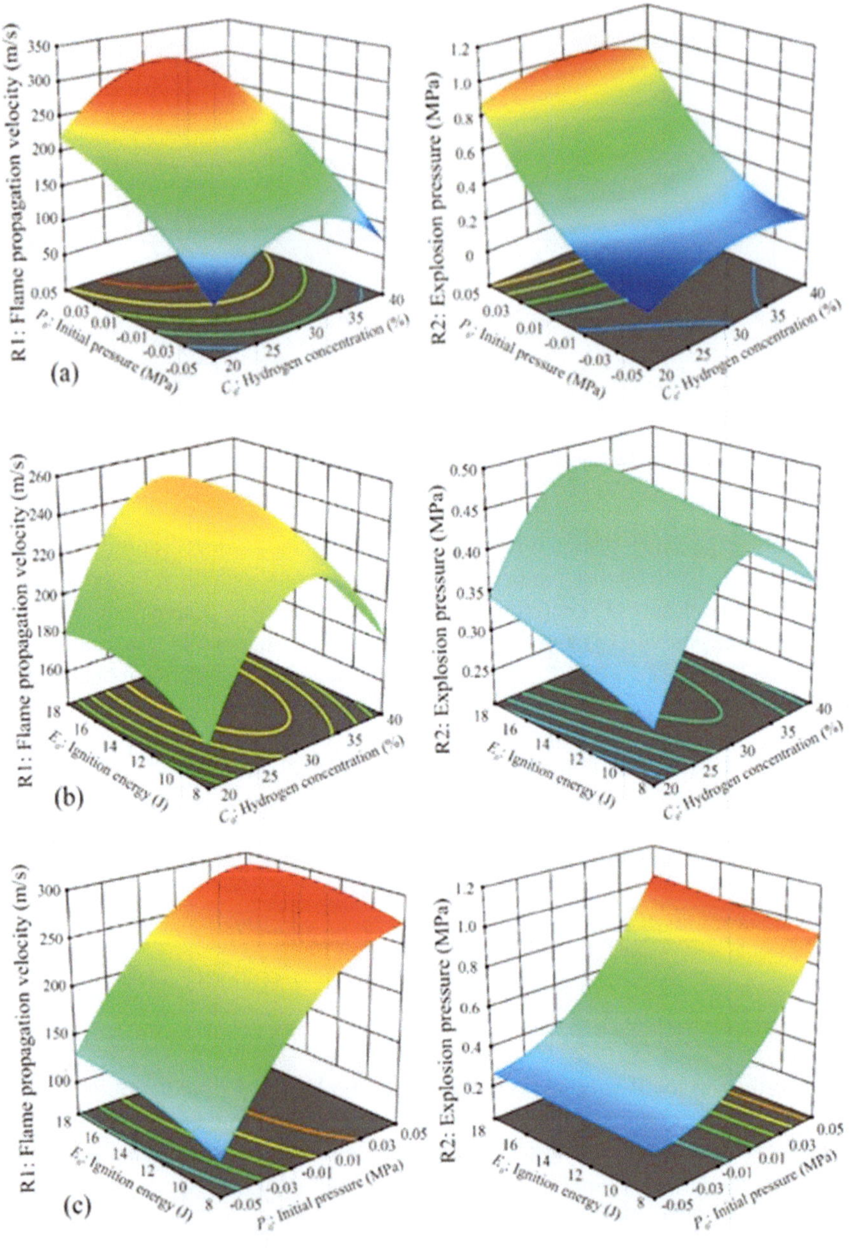

Fig. 11.11 The response surface plot for interaction effects of **a** C_0 and P_0, **b** C_0 and E_0, **c** P_0 and E_0 on the V_f and P_f

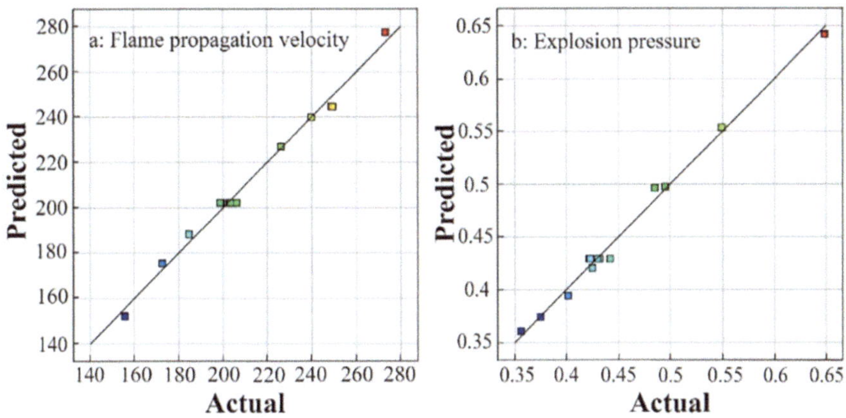

Fig. 11.12 The predicted versus actual values plot for V_f and P_f

Figure 11.13 presents the response surface plots for the interaction effects of α and H on V_f and P_f. With the decrease of α and H, the V_f value rises significantly, and the α has a greater effect on V_f model than H, as shown in Fig. 11.13a. The P_f value increases significantly with increasing the H and the decreasing theα. The H has a greater effect on the P_f model than α, as shown in Fig. 11.13b.

In summary, the changes of the initial conditions will cause the changes of the V_f and P_f when the flame arrester structure parameters values are fixed. However, the critical flame propagation velocity (V_c) and explosion pressure (P_c) at which the flame arrester can successfully quench flame would not change with changing the initial conditions. For example, when the flame arrester porosity and element thickness are 0.35 and 80 mm respectively, the V_c and P_c of the flame arrester under

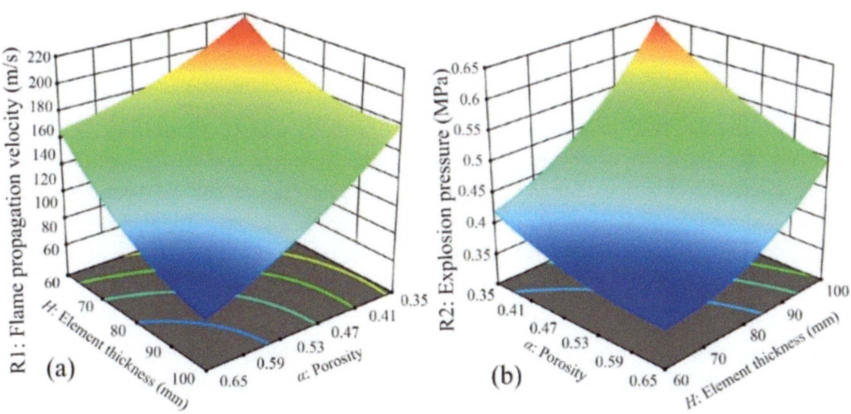

Fig. 11.13 Response surface plot for interaction effects of different structure parameters on V_f and P_f

different initial conditions are 183 m/s and 0.51 MPa respectively. However, when the initial conditions values are fixed, the changes in the structure parameters would cause changes in the V_c and P_c values. For example, when the porosity (0.35–0.65) and element thickness (60–100 mm) of the flame arresters are changed, the V_c value increases from 65 to 184 m/s, and the P_c value increases from 0.15 to 0.61 MPa. Therefore, it can be concluded that the structure parameters of flame arresters are important influence factors for V_c and P_c.

Critical Flame Quenching Criterion Model

The above analysis shows that The V_c and P_c could be used as important parameters to measure the flame quenching performance of flame arrester, and only related to the structure parameters of flame arrester, so the V_c and P_c prediction models under different structure parameters are established by RSM respectively. The interaction effects of porosity and element thickness on the flame quenching performance are studied by RSM. The experimental design scheme is shown in Table 11.4. The response parameters are the critical flame propagation velocity (V_c) and explosion pressure (P_c), and the design parameters are the flame arrester porosity (α) and the element thickness (H). Meanwhile, the two-factor prediction models of V_c and P_c under different structure parameters are obtained by RSM as follows:

$$\begin{aligned} V_c &= 14.19657 + 297.39382 * \alpha + 2.62025 * H \\ &\quad - 2.02827 * \alpha H - 478.59893 * \alpha^2 - 0.004836 * H^2 \\ R^2 &= 0.9898 \end{aligned} \tag{11.5}$$

$$\begin{aligned} P_c &= -0.255931 + 0.710307 * \alpha + 0.013633 * H \\ &\quad - 0.016833 * \alpha H - 0.431418 * \alpha^2 - 0.000011 * H^2 \\ R^2 &= 0.9868 \end{aligned} \tag{11.6}$$

According to the P-values corresponding to the V_c and P_c prediction models in ANOVA, and the comparison results between the predicted and experimental values in Fig. 11.14, as well as the R^2 value of the equations, it can be found that the prediction models of V_c and P_c have good accuracy with experimental results. The P-values of structure parameters are < 0.05 according to the ANOVA results, which indicates that α and H have a significant influence on V_c and P_c. By comparing the F-values corresponding to the design parameters, it can be found that the porosity (α) and have a greater effect on the V_c and P_c models than element thickness (H).

Figure 11.15 presents the response surface plots for the interaction effects of α and H on the V_c and P_c. It can be seen that both V_c and P_c values increase significantly with decreasing porosity and increasing element thickness, and the influence of porosity on the V_c and P_c values is more significant compared to the

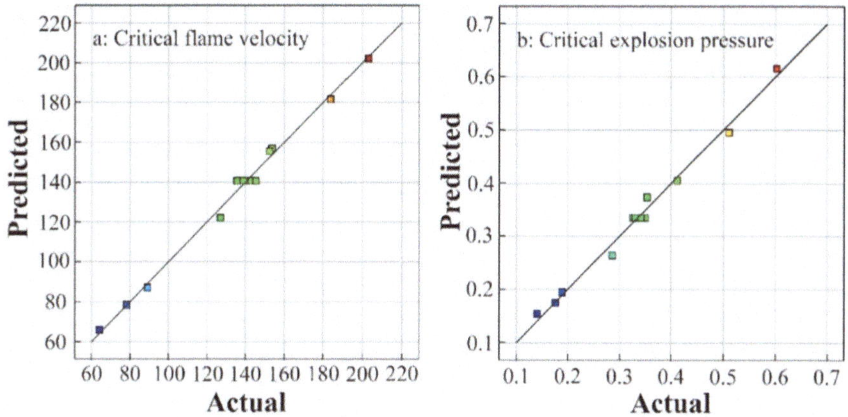

Fig. 11.14 The predicted versus actual values plot for V_c and P_c

element thickness. Therefore, in order to improve the flame quenching performance, the design process of flame arrester should focus on the influence of porosity.

From the above analysis, the two prediction models of critical flame propagation velocity (V_c) and explosion pressure (P_c) can effectively reflect the flame quenching performance of flame arrester. Meanwhile, the critical flame quenching criterion can be proposed based on the multi-factor prediction models of flame propagation velocity (V_f) and explosion pressure (P_f) entering the flame arrester, and the two-factor prediction models of V_c and P_c. When the V_f is greater than the V_c, the flame quenching failure; conversely, the flame quenching success.

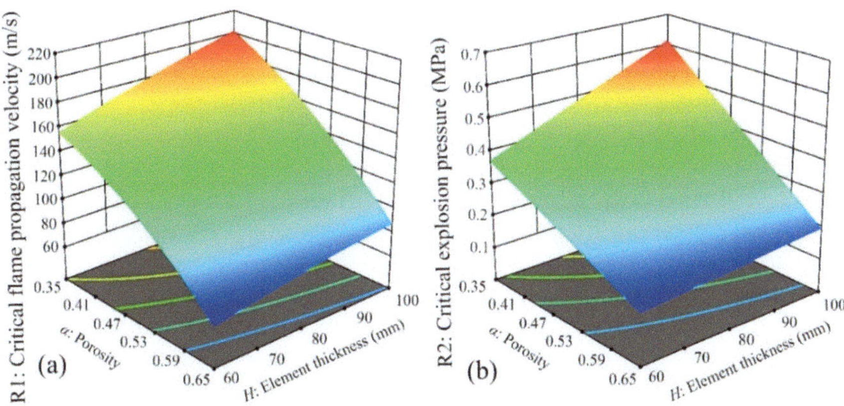

Fig. 11.15 The response surface plot for interaction effects of different structure parameters on the V_c and P_c

11.3 Conclusions

Based on the self-built experimental apparatus, the effects of initial conditions, structure parameters, and expansion chamber structures on the flame quenching performance parameters (flame propagation velocity (V_f) and explosion pressure (P_f) entering the flame arrester) were investigated. Then the flame quenching mechanism of the crimped-ribbon flame arrester and the effect of influencing factors on the flame quenching process were analyzed. Meanwhile, multi-factor prediction models of V_f and P_f under different structure parameters and initial conditions were established based on the RSM respectively. Besides, the multi-factor prediction models of the critical flame propagation velocity and explosion pressure under different structure parameters were established respectively by RSM, and the critical flame quenching criterion was proposed. The conclusions could be summarized as follows:

(1) With the increase of hydrogen concentration, the V_f and P_f showed a variation trend of first increasing and then decreasing. The V_f and P_f were gradually increased with increasing the initial pressure and ignition energy, and the flame quenching failure probability also was significantly increased. Meanwhile, the decrease in porosity and the increase in element thickness could lead to a significant increase in the flame quenching success probability. Then, the extended type and baffle type expansion chambers could efficiently quench the hydrogen explosion flame compared with the conventional type expansion chamber, and the extended type expansion chamber had a better flame quenching performance. Meanwhile, the initial conditions could affect the V_f and P_f, which in turn affect the heat transfer and quenching process inside the narrow channels. The flame was dispersed into the smaller flame structure and experienced a longer propagation distance with increasing the element thickness and decreasing the porosity. This could enhance the heat transfer between the explosion flame and the wall surface, and the probability of collision destruction between the free radicals and the wall, resulting in the quenching of the explosion flame. Besides, the expansion chamber structure could effectively reduce the flame propagation velocity and explosion pressure entering the narrow channels, which could improve the quenching flame performance.

(2) Based on the RSM, the multi-factor prediction models of V_f and P_f under different initial conditions were established respectively. The influence order of a single initial condition on the V_f and P_f by the ANOVA of the models was: $P_0 > C_0 > E_0$. The influence order of the interaction effects on V_f and P_f was: (C_0 and P_0) > (P_0 and E_0) > (C_0 and E_0). Meanwhile, the two-factor prediction models of V_f and P_f under different structure parameters were obtained respectively. It can be found that the α had a greater effect on V_f model, and H had a greater effect on P_f model. Besides, the V_c and P_c were only related to the structure parameters of flame arrester.

(3) Based on the RSM, the multi-factor prediction models of V_c and P_c under different structure parameters were established respectively. And the porosity (α) had a greater effect on the V_c and P_c models than the element thickness (H)

by the ANOVA of the models. Combining the multi-factor prediction models of V_f and P_f, the critical flame quenching criterion was proposed. It provided an accurate prediction method for the flame quenching success of flame arrester under different working conditions.

For practical application, to improve flame quenching performance of hydrogen crimped-ribbon flame arrester and achieve optimal design of flame arrester, three suggestions are proposed based on the above works: (1) It should be reducing porosity and increasing element thickness as far as possible within the acceptable pressure drop range, and focusing on the effect of porosity during the optimal design of flame arrester; (2) Under the condition of ensuring normal gas transmission, the delivery pressure in the pipe should be reduced as much as possible; (3) The expansion chamber structure should be optimized to reduce the flame propagation velocity and explosion pressure entering the flame arrester as much as possible.

References

1. Lin CD, Cao XY, Wang ZR, Wei JS, Xu JJ. Research on quenching performance and multi-factor influence law of hydrogen crimped-ribbon flame arrester using response surface methodology. Fuel. 2022;326:124911.
2. Le AT, Tan Z-H, Sivakumar R, Pung S-Y. Predicting the photocatalytic performance of metal/metal oxide coupled TiO_2 particles using Response Surface Methodology (RSM). Mater Chem Phys. 2021;269:124739.
3. Kim W, Mogi T, Dobashi R. Effect of propagation behaviour of expanding spherical flames on the blast wave generated during unconfined gas explosions. Fuel. 2014;128:396–403.
4. Sun S, Liu G, Liu J, Ye C, Ren J, Bi M. Effect of porosity and element thickness on flame quenching for in-line crimped-ribbon flame arresters. J Loss Prevent Proc. 2017;50:221–8.
5. Heidari A, Wen JX. Flame acceleration and transition from deflagration to detonation in hydrogen explosions. Int J Hydrogen Energy. 2014;39(11):6184–200.
6. Thomas G, Oakley G, Bambrey R. Fundamental studies of explosion arrester mitigation mechanisms. Process Saf Environ. 2020;137:15–33.
7. Jin K, Duan Q, Chen J, Liew KM, Gong L, Sun J. Experimental study on the influence of multi-layer wire mesh on dynamics of premixed hydrogen-air flame propagation in a closed duct. Int J Hydrogen Energy. 2017;42(21):14809–20.
8. Li Q, Lu S, Wang C. Effect of orifice plates spaces on flame propagation in a square cross-section channel. Int J Hydrogen Energy. 2019;44(40):22537–46.
9. Saiki Y, Suzuki Y. Effect of wall surface reaction on a methane-air premixed flame in narrow channels with different wall materials. Proc Combust Inst. 2013;34(2):3395–402.
10. Chu J, Liu Y, Lv X, Li Q, Dong H, Song Y, et al. Experimental investigation on blockage predictions in gas pipelines using the pressure pulse wave method. Energy. 2021;230:120897.
11. Pereira LMS, Milan TM, Tapia-Blacido DR. Using Response Surface Methodology (RSM) to optimize 2G bioethanol production: a review. Biomass Bioenergy. 2021;151:106166.

Chapter 12
Multi-factor Influencing on Detonation Quenching Performance of Crimped-Ribbon Flame Arrester

12.1 Experimental Apparatus and Procedures

In the previous chapter, the graphic illustration of the experimental apparatus was provided. In this case, the lengths of detonation and protected pipes were 4000 mm and 2000 mm separately. The flame arrester (FA) equipment consisted of flame arrester element (CF-E) and expansion chamber (EC). In the experiment, the CF-E with different porosities and element thicknesses were used. Then, different EC (the conventional, baffle, and extended types) were adopted to connect with the CF-E. There were three types of EC with the same volume and their only difference was in the internal structure, as shown in Fig. 11.2. In contrast to the conventional EC, the baffle-type EC had an internal annular baffle (inside diameter: 70 mm), which was 30 mm and 240 mm away from the CF-E and the flange end face. Besides, the tube with 155 mm was mounted in the extended EC and the circular hole with 12 mm were evenly distributed along the circumferential direction of tube port [1].

Meanwhile, distribution equipment (RCS2000-B, accuracy: 1%) was adopted to prepare the premixed gas. The energy range of high-voltage ignition system was 1–20 J. The pressure transducers (CY400, accuracy: 0.25%) were mounted on the detonation pipe, while the mounting positions were 400 and 200 mm away from the FA separately. Flame detectors (CKG100, No. 1–4) were mounted on the detonation pipe. No. 5 and No. 6 flame detectors were mounted on the EC and protected pipe separately. Then, the detonation velocity inside the EC and FA-E was calculated using the response interval time and spacing of No. 4–6 flame detectors, which was used to analyze and evaluate the attenuating and quenching performances of the FA.

Besides, the initial conditions (P_0, C_0 and E_0) were regulated through the gas distribution system and ignition system, while the element thickness (H) and porosity (α) of FA were adjusted by replacing the FA-E. The experiment program for the influences of initial condition and structure parameter of FA-E was shown in Table 12.1. Before the experiment, the air inside the pipe was exhausted to the outside using a vacuum pump. Then, the gas distribution system continuously injected hydrogen/air

Z. Wang and X. Cao, *Flame Resistance Characteristics and Its Influence Mechanism*,
https://doi.org/10.1007/978-981-95-6119-3_12

premixed gas into the pipe until the vacuum gauge reached a standard atmospheric pressure of 0.1 MPa.

To guarantee the precision of the experiment results, all experiments were repeated five times. The maximum deviation of the flame velocity and pressure was less than 5.2% and 7.2% respectively. To further analyze the effects of influencing factor and its multi-factor interaction on explosion parameters, the response surface method (RSM) with good prediction performance and accuracy was used [2]. The influences of multiple factors on flame velocity and pressure entering the FA under the action of initial conditions were investigated by RSM, and multi-factors prediction models were established. Then, the Box-Behnken method of RSM was adopted, and the experimental program was presented in Table 12.2.

Table 12.1 Experiment program for the influences of initial condition and structure parameter

Initial conditions					Structure parameters				
No.	Influence factors		Initial condition	Structure parameter	No.	Influence factors		Structure parameter	Initial condition
A1	C_0	20%	$P_0 = 0$ MPa $E_0 = 13$ J	$\alpha = 0.35$	D1	α	0.35	$H = 80$ mm	$C_0 = 30\%$ $P_0 = 0$ MPa $E_0 = 13$ J
A2		25%			D2		0.5		
A3		30%			D3		0.65		
A4		35%							
A5		40%							
B1	P_0	0 MPa	$C_0 = 30\%$ $E_0 = 13$ J	$H = 80$ mm	E1	H	60 mm	$\alpha = 0.5$	
B2		0.01 MPa			E2		80 mm		
B3		0.02 MPa			E3		100 mm		
B4		0.03 MPa							
B5		0.04 MPa							
C1	E_0	3 J	$C_0 = 30\%$ $P_0 = 0$ MPa						
C2		8 J							
C3		13 J							
C4		18 J							

Table 12.2 Experiment program of RSM

Initial conditions							
No.	C_0 (%)	P_0 (MPa)	E_0 (J)	No.	C_0 (%)	P_0 (MPa)	E_0 (J)
1	20	0	13	10	30	0.04	8
2	40	0	13	11	30	0	18
3	20	0.04	13	12	30	40	18
4	40	0.04	13	13	30	0.02	13
5	20	0.02	8	14	30	0.02	13
6	40	0.02	8	15	30	0.02	13
7	20	0.02	18	16	30	0.02	13
8	40	0.02	18	17	30	0.02	13
9	30	0	8				

12.2 Results and Discussions

12.2.1 *Influencing Factors on Quenching Performance*

Flame Propagation Characteristics

Figure 12.1 gives the flame signals at different positions. According to the distance between adjacent flame detectors ($L_1 = 200$ mm, $L_2 = 500$ mm, $L_3 = 200$ mm, $L_4 = 320$ mm, and $L_5 = 590$ mm) and the interval time between flame signals, the flame velocity at different zones is calculated. It could be concluded that the flame velocity gradually increases as the distance from the left inlet of the FA decreases. Meanwhile, the critical detonation velocity (V_{CJ}) with a hydrogen concentration of 30% is calculated to be 1973 m/s by CEA (NASA computer program) [3].

It is found that the flame velocity (V_3) value is greater than the V_{CJ} value, indicating that the explosion flame reaches the state of detonation. According to the flame velocity (V_4) in the zone of No. 4–5 flame detectors, it could be inferred that the flame velocity is decreased inside the EC compared to the velocity (V_3) accessing the FA, and the V_4 is slightly lower than V_{CJ}. This presents that the EC can availably attenuate the flame velocity, thereby weakening the detonation intensity acting on the flame arrester element (FA-E). Based on the interval time and distance between the No. 5–6 flame detectors, the flame velocity (V_5) inside the FA-E is calculated. It is found that the flame velocity is further decreased to lower than the V_{CJ} value due to the attenuation effect of the FA-E. This means that the flame arrester (the EC and FA-E) can effectively attenuate the intensity of the detonation, causing the detonation wave to be decoupled from the flame front and decayed into the deflagration state.

Figure 12.2 presents the pressure curves at different positions. The pressure peak value gradually is improved as the explosion flame accelerates and propagates. The pressure curves suddenly rise as the pressure wave enters the monitoring zone, and

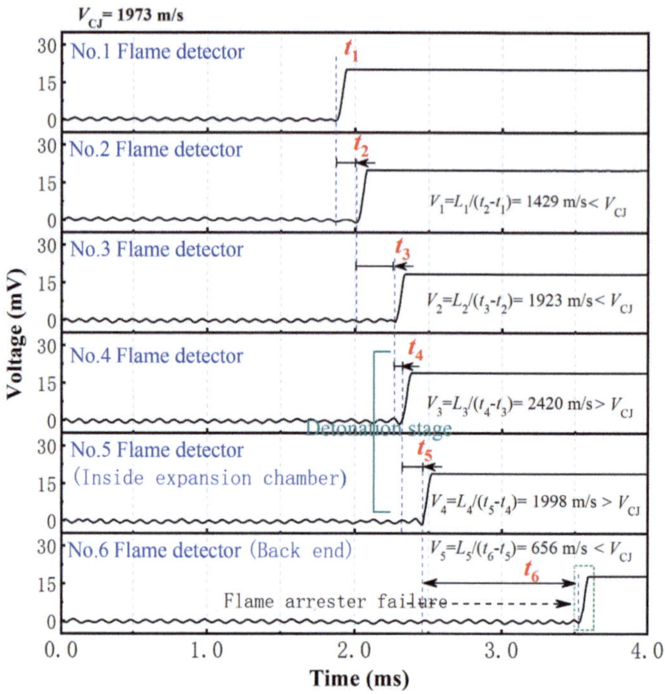

Fig. 12.1 Flame signals at different positions

then the curves oscillate and reduce due to the action of Taylor rarefaction wave. And the oscillation periods of the curves (No. 1 and 2) during the reducing stage are 0.116 ms and 0.187 ms, respectively. Then, the flame detector (No. 4) and the pressure transducer (No. 2) are mounted on the inlet zone near the FA, the values measured at the zone are taken to be the flame velocity (V_f) and pressure (P_f) accessing FA. Meanwhile, the pressure value ($P_{No.2} = 1.93$ MPa) measured by No. 2 pressure transducer is significantly greater than the CJ pressure value ($P_{CJ} = 1.56$ MPa). Combining with the flame velocity at the front of the FA, it indicates that the explosion flame entering the FA reaches the detonation state [3].

Effects of Initial Conditions

Figure 12.3a–c present the V_f, P_f and the flame velocity (V_5) inside FA-E under different hydrogen concentrations, initial pressures and ignition energies, respectively. Figure 12.3a demonstrates that both the V_f and P_f exhibit a tendency to increase and then reduce as the hydrogen concentration rises. However, the velocity of the flame is significantly reduced as it propagates into the FA, and the attenuation extent of velocity presents a trend of reducing and then rising. Both the V_f and P_f reach the maximum value ($V_f = 2420$ m/s; $P_f = 1.93$ MPa) at the concentration

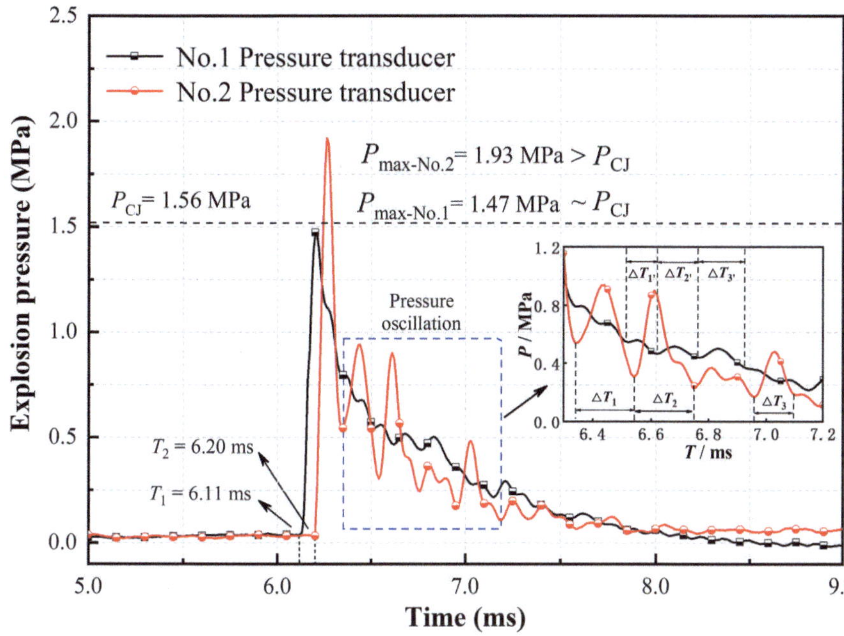

Fig. 12.2 Pressure curves at different positions

($C_0 = 30\%$) close to the stoichiometric ratio. That phenomenon is similar to the one that Li and Wang mentioned. As the hydrogen concentration is below the equivalent concentration, its increase could enhance the quantity of free radicals during the chain reaction. It results in an increase of the chemical reaction rate and a large amount of heat is released. The detonation strength increases continuously and reaches the maximum at the equivalent ratio. However, as the hydrogen concentration is greater than the equivalent concentration, the thermal decomposition of excess hydrogen will absorb an amount of heat with the increase of hydrogen concentration. Meanwhile, relatively excess hydrogen cannot completely react with oxygen, leading to a reduction in the chemical reaction rate. Then the V_5 is reduced to 656.5 m/s and lower than the 1/2 V_{CJ} value, indicating that the detonation is attenuated and decoupled into the deflagration flame. Compared with the V_f value, the attenuation extent of V_6 is 72.8%. The V_f and P_f are significantly decreased as the hydrogen concentration is far away from 30%. And the V_5 is significantly decreased, leading to the attenuation effect of the FA being improved. Figure 12.3b shows the increase in initial pressure causes a significant improvement in both V_f and P_f. The hydrogen content and energy density per unit volume are significantly improved due to the increase of the initial pressure. This will release more energy from the reaction and lead to a reduction in the detonation cell size, further causing an increase in detonation intensity [4]. In particular, V_f and P_f reach the peak value ($V_f = 2904$ m/s; $P_f = 2.68$ MPa) respectively at P_0 of 0.04 MPa. But the attenuation effect of FA is weakened, and the attenuation extent of flame velocity gradually is decreases (from 72.8 to 70.7%).

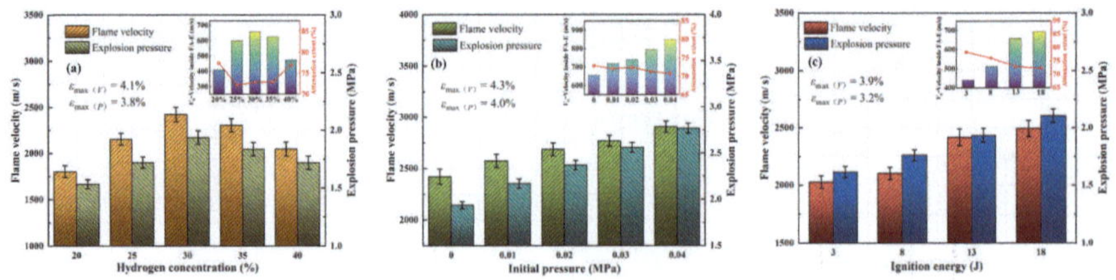

Fig. 12.3 The V_f, P_f and quenching effects under the action of initial conditions

Besides, Fig. 12.3c presents that both V_f and P_f values are raised by increasing initial energy, the values reach the maximum value at an ignition energy of 18 J. The chain-reaction energy and the free radicals number are raised by the enhancement of ignition energy, causing the enhancement of flame velocity and pressure [4]. The V_6 inside the FA also presented an increase change trend as the ignition energy rise, indicating the obvious decrease in the attenuation extent of flame velocity (from 78.3 to 72.2%). In summary, the V_f and P_f are significant influencing factors on the quenching performance of the FA, and the initial conditions have an indirect effect on the quenching performance by immediately influencing the V_f and P_f.

Effects of Expansion Chamber Structures

Figure 12.4a, b show the propagation velocity (V_4), attenuation extent and quenching results of the detonation in the three type EC. It is obvious from Fig. 12.4a that the flame velocity entering the FA is significantly attenuated by the three ECs, which indirectly affects the quenching effect by reducing the detonation intensity acting on the FA-E [5]. The attenuation effect of EC on flame velocity tends to decrease and then increase as the hydrogen concentration rises (20–40%). The attenuation effect of EC on detonation velocity is the lowest at a hydrogen concentration of 30%, which causes the flame velocity to exceed the quenching limit value of FA-E. And this leads to the failure of FA using the conventional and baffle type ECs. However, the attenuation effect of EC on flame velocity is enhanced due to the weaker detonation intensity as hydrogen concentration is far away from 30%. And the flame velocity entering the FA-E is lower than the V_{CJ} value, which leads to the successful quenching of detonation by FA-E. Especially, the flame velocity is reduced to less than 1/2 V_{CJ} value under the action of the extended-type EC, which causes the detonation to decouple into the deflagration flame and greatly enhances the quenching effect of the FA-E. Meanwhile, the propagation process of the detonation is significantly affected by the structure of EC, which further affects the attenuation performance of the EC. The experimental results show that the baffle and extended type ECs are more effective than the conventional type EC in reducing the detonation velocity entering the FA-E. The influence order of the ECs on the detonation velocity is: the extended-type EC > baffle-type EC > conventional-type EC. The detonation entering the baffle-type EC can be diffused and collided with annular baffle, causing

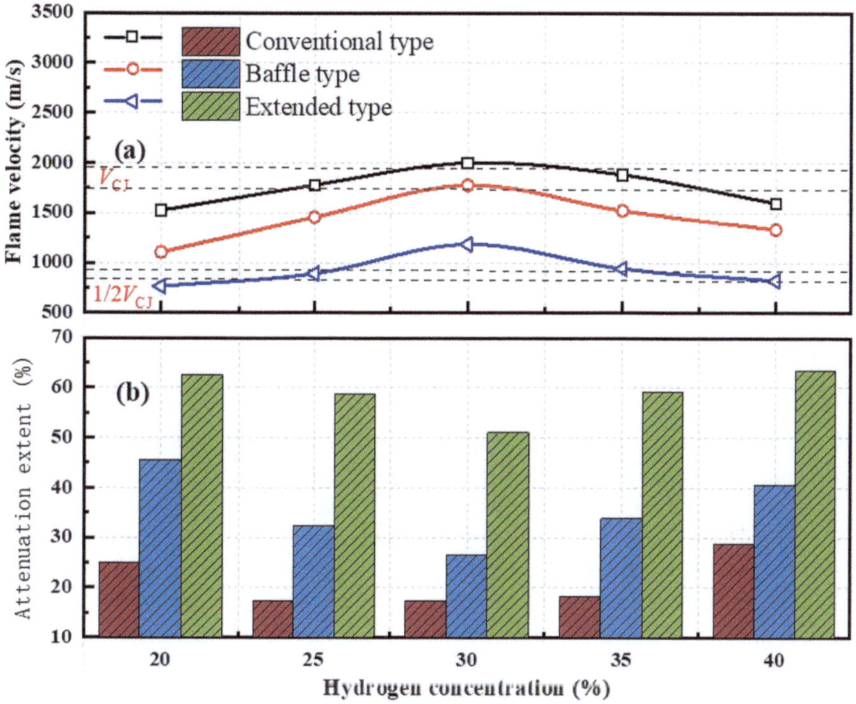

Fig. 12.4 The flame velocity and quenching results under the action of expansion chamber (√—flame quenched, ×—flame unquenched)

the losses of momentum and energy. Then, the reflected wave generated by collision effect act on detonation wave, which causes the loss of transverse wave and the reduction of the reaction rate. And this brings about the attenuation of the detonation velocity. Furthermore, the flame propagates radially and interacts with the chamber wall to produce energy loss when it reaches the holes in the extended EC [6]. The reflected wave was formed after the detonation wave of axially propagation collides with the FA-E. The reflected wave is destroyed after multiple reflection propagation due to the blocking effect of extended pipe and the chamber walls, leading to an obvious decrease in the detonation intensity acting on the FA-E.

Effects of Structure Parameters

Figure 12.5a, b show the flame velocity (V_5) inside FA-E, and its attenuation extent at different porosity and element thickness. The flame velocity in FA-E gradually is decreased with decreasing porosity in Fig. 12.5a, while the attenuation extent is gradually increased (from 53.9 to 67.1%). Compared to flame velocity (V_4) inside the EC, flame velocity is decreased significantly after entering the FA-E and is less than

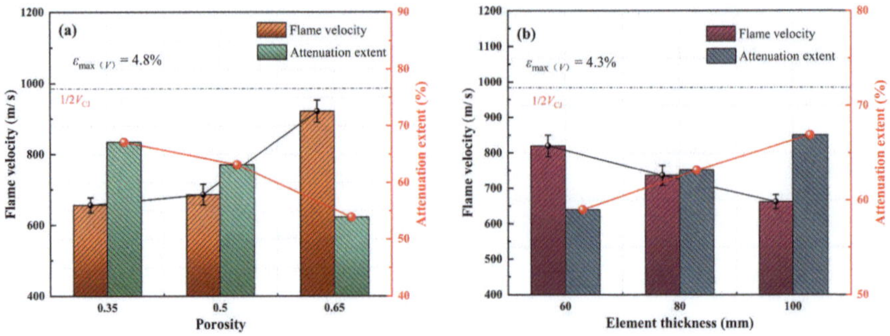

Fig. 12.5 The quenching effects under the action of structure parameters

1/2 V_{CJ} value. Especially, the V_6 is reduced to 656 m/s as the porosity is 0.35, and the attenuation extent of velocity is increased to 67.1%. The collision effect between the detonation and the FA-E is enhanced due to the decrease in porosity. And the momentum loss and the velocity deficit is also significantly raised, which brings about the reduction of detonation velocity entering the narrow channel. Then, the flame can be divided into tinier flame as the porosity decreases, which will enhance the heat transfer of wall and the destruction of free radicals. And the velocity deficit is increased under action of the boundary layer and viscous wall effects, which enhances the attenuation effect of FA-E. Meanwhile, the flame velocity in the FA-E will be gradually decreased as the element thickness increases in Fig. 12.5b, and the attenuation velocity extent is gradually improved. The flame velocity is reduced to 661.6 m/s at an element thickness of FA-E of 100 mm. And the attenuation extent of FA-E on flame velocity is gradually increased (from 58.9 to 66.8%) as the element thickness increases. The enhancement in the element thickness extends the action time between the flame and channel wall, which further enhances attenuation effect such as the wall heat transfer and free radical destruction. Hence, the detonation flame produces a large velocity deficit and decouples into deflagration flame, which leads to an improvement in the quenching performance of the FA-E. It can be found that variation in the structure parameters can influence the propagation and extinction process of detonation, thus influencing the quenching performance of the FA, and the effect of porosity on quenching performance is greater than that of the element thickness [7].

12.2.2 Interactions of Initial Conditions on Quenching Performance

As a close coupling between the shock wave and the flame surface, detonation is a process of instantaneous chemical reaction with low entropy increase. The adiabatic compression of the shock wave makes the gas reach the ignition temperature, which causes a violent chemical reaction. As a strong compression wave, there is a significant increase in pressure and density after the detonation wave surface [8]. Due to the supersonic propagation characteristic of the detonation wave and the flame surface, the unburned gas in the pipe is not affected and remains in the initial state before the detonation wave arrives. Different from the detonation structure, the deflagration consists of a precursor pressure wave and a flame surface immediately following it. In the process of deflagration propagation, the mass and heat transfer in the reaction zone cause the gas to reach the ignition temperature and trigger the chemical reaction to realize the deflagration flame propagation [9]. Then, the reflected wave generated by the FA-E can disturb the flame front and change its reaction rate due to the low subsonic propagation of deflagration wave. This also causes the deflagration flame velocity and pressure value entering the FA not only depend on the initial conditions, but also on the structural parameters of the FA-E and the EC [7]. Meanwhile, it is found that the V_f and P_f entering the FA are affected significantly with the change of the structural parameters of the FA-E in the previous experiment of the deflagration quenching characteristics of the CFA [1]. And the flame velocity and pressure are decreased by 26.5% and 22.4% respectively as the porosity increased from 0.35 to 0.65. However, the unburned gas behind the shock wave is not disturbed and will not form a reflected wave acting on the flame surface due to the propagation characteristics of detonation. Therefore, the detonation velocity and pressure value entering the FA are not affected by the structural parameters of the FA-E. Besides, it is found that detonation velocity and pressure value entering the FA do not change significantly with the change of the structural parameters of the FA in this research. And the flame velocity and pressure value are 2420 m/s and 1.93 MPa under different flame arrester structure parameters ($P_0 = 0$ MPa, $C_0 = 30\%$ and $E_0 = 13$ J). This further presents that the structural parameters of the FA-E do not affect the detonation velocity and pressure value entering the FA. However, the detonation reaction process is significantly affected by initial conditions. The change of C_0 and P_0 will significantly affect the energy density and detonation sensitivity of the gas [4]. And the change of E_0 will significantly affect the energy of the chain reaction and the number of free radicals, which will affect the reaction rate and then affect the flame velocity and pressure value [10]. Therefore, this chapter studies the influence of initial conditions and their multi-factor interaction on the detonation velocity and pressure entering the FA and establishes the prediction models.

Effect of Single Factor

The multi-factor interactions of initial conditions on the explosion parameters are investigated by RSM based on the single-factor effects. Then BBD experimental design with three factors and three levels is used, and the research program is shown in Table 12.2. Detonation velocity (V_f) and the pressure (P_f) accessing FA are the response parameters, while the initial pressure (P_0), hydrogen concentration (C_0) and ignition energy (E_0) are the design parameters. Moreover, multi-factor prediction models are established separately as follows:

$$
\begin{aligned}
V_f &= -2430.5872 + 264.9766 * C_0 + 35,063.5750 * P_0 + 33.26187 * E_0 \\
&\quad - 423.5875 * C_0P_0 - 0.3080 * C_0E_0 - 564.1750 * P_0E_0 - 3.9953 * C_0^2 \\
&\quad - 26888.75 * P_0^2 + 0.4868 * E_0^2 \\
R^2 &= 0.9923
\end{aligned} \tag{12.1}
$$

$$
\begin{aligned}
P_f &= -2.6469 + 0.2714 * C_0 + 13.6156 * P_0 + 0.0484 * E_0 + 0.2261 * C_0P_0 \\
&\quad + 0.5500E - 03 * C_0E_0 - 0.4370 * P_0E_0 - 0.4503E - 02 * C_0^2 \\
&\quad + 24.7625 * P_0^2 - 0.1456E - 02 * E_0^2 \\
R^2 &= 0.9952
\end{aligned} \tag{12.2}
$$

To check the correctness of prediction models and confirm the influences of different factors and their interactions on the V_f and P_f, the variance analysis (ANOVA) is adopted [4]. Tables 12.3 and 12.4 show the results of the ANOVA for the prediction models. It is evident from ANOVA that the error probability (P-value) of the models is smaller than 1E–04, and the models are in "significant" state and their "lack of fit" is in "not significant" state [3]. Furthermore, the R^2 values of both models are above 0.90.

Besides, Fig. 12.6 depicts that the predicted values of the models are well consistent with experimental values. In conclusion, the prediction models demonstrate good accuracy and effectiveness [11]. According to the ANOVA, The P-values of P_0, C_0 and E_0 are lower than 0.05, indicating that the factors have an observably effect on

Table 12.3 ANOVA for the predicted model of the V_f

Source	Sum of squares	Df	Mean Square	F-value	P-value	
Model	1.60E + 06	9	1.78E + 05	99.92	< 0.0001	Significant
C_0	1.31E + 05	1	1.31E + 05	73.36	< 0.0001	
P_0	6.22E + 05	1	6.22E + 05	349.52	< 0.0001	
E_0	1.29E + 05	1	1.29E + 05	72.45	< 0.0001	
C_0P_0	28,708.22	1	28,708.22	16.12	0.0051	
C_0E_0	948.64	1	948.64	0.5327	0.4892	
P_0E_0	12,731.74	1	12,731.74	7.15	0.0318	
$C_0{}^2$	6.72E + 05	1	6.72E + 05	377.45	< 0.0001	
$P_0{}^2$	487.08	1	487.08	0.2735	0.6171	
$E_0{}^2$	623.82	1	623.82	0.3503	0.5725	
Residual	12,464.84	7	1780.69			
Lack of fit	9827.74	3	3275.91	4.97	0.0777	Not significant
Pure error	2637.1	4	659.28			
Cor total	1.61E + 06	16				

Table 12.4 ANOVA for the predicted model of the $\boldsymbol{P_f}$

Source	Sum of squares	df	Mean square	F-value	P-value	
Model	1.88	9	0.2089	161.62	< 0.0001	Significant
C_0	0.1344	1	0.1344	104.01	< 0.0001	
P_0	0.7896	1	0.7896	610.82	< 0.0001	
E_0	0.0673	1	0.0673	52.1	0.0002	
C_0P_0	0.0082	1	0.0082	6.33	0.0401	
C_0E_0	0.003	1	0.003	2.34	0.1699	
P_0E_0	0.0076	1	0.0076	5.91	0.0453	
$C_0{}^2$	0.8536	1	0.8536	660.34	< 0.0001	
$P_0{}^2$	0.0004	1	0.0004	0.3196	0.5895	
$E_0{}^2$	0.0056	1	0.0056	4.32	0.0764	
Residual	0.009	7	0.0013			
Lack of fit	0.0075	3	0.0025	6.48	0.0513	Not significant
Pure error	0.0015	4	0.0004			
Cor total	1.89	16				

V_f and P_f. Considering that the P-values of P_0, C_0 and E_0 are all less than 1E−04, and a larger F-value of the influence factor indicates that it has a larger impact on V_f and P_f, so the influence order of the three factors on V_f and P_f is: $P_0 > C_0 > E_0$.

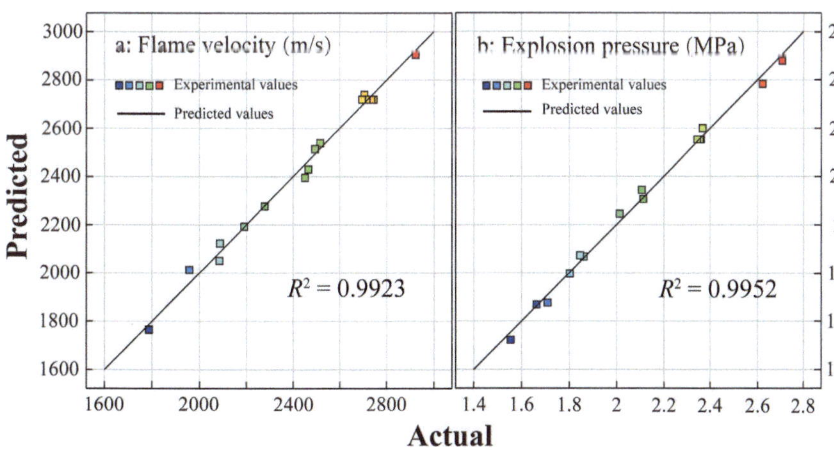

Fig. 12.6 The predicted versus actual values plot using prediction models

Effect of Multi Factors

Based on the ANOVA, it is found that the two-factor interactions of P_0, C_0 and E_0 have observably influence on the prediction models. Figure 12.7a–c demonstrate the diagrams of the two-factor interaction on V_f and P_f, respectively. Based on the interaction diagrams of C_0 and P_0 in Fig. 12.7a, it is found that the slope of the surface with increase in P_0 is significantly greater than that with changes in C_0, indicating that the effect of P_0 on V_f and P_f is greater than that of C_0. The interaction diagram of C_0 and E_0 in Fig. 12.7b also presents that the influence of C_0 on V_f and P_f is greater than that of E_0, which is consistent with the results of ANOVA. Meanwhile, the V_f and P_f values under the interaction of C_0 and E_0 are significantly lower than those of the other two interactions, and the P-value of the interaction of C_0 and E_0 in the ANOVA is greater than 0.05, which indicates that the interaction of C_0 and E_0 has a weak impact on V_f and P_f values. In comparison of the interaction between P_0 and E_0 in Fig. 12.7c, the contours under the interaction of C_0 and P_0 have an elliptical shape, indicating that the interaction between the two factors has a better impact on the V_f and P_f. Furthermore, the order of the two factor interaction effects on V_f and P_f is: (C_0 and P_0) > (P_0 and E_0) > (C_0 and E_0) by contrasting the P-values of the two-factor interactions in the ANOVA.

12.2.3 Analysis of Quenching Mechanism

Figure 12.8 shows the propagation and quenching process of the detonation inside the EC and the narrow channel. The flame velocity and pressure reach the detonation state as the flame propagates to the inlet of the FA. The detonation intensity is reduced by the increase in cross-sectional area and the rarefaction wave effect inside the EC, and causing the enhancement of the induction zone distance between the detonation wave and flame front. Meanwhile, the energy loss and the velocity deficit are raised due to the diffraction and collision effects between the detonation and EC wall, which causes the reduction in the velocity entering the FA-E [12]. In addition, the structure of the EC will affect the propagation and attenuation characteristics of the detonation inside the EC. And the diffraction and the collision effects of the chamber wall are enhanced under action of the baffle and extended type ECs [7]. This will lead to a rise in the energy loss and velocity deficit of the detonation, thereby reducing the detonation intensity applied to FA-E. As a result, the flame is more easily extinguished by the narrow channel due to the attenuation effect of EC.

Meanwhile, mass dissipation and momentum loss occur under the action of the boundary layer and the viscous wall effects after the detonation entered FA-E. This causes an increase in the velocity deficit and the induction zone distance, ultimately leading to the attenuation and decoupling of the detonation. The detonation flame is divided into a lot of tiny flames after entering the FA-E, and the energy loss is enhanced under the action of the wall endothermic. This leads to a reduction in the reaction rate and flame temperature, and the flame is successfully quenched as

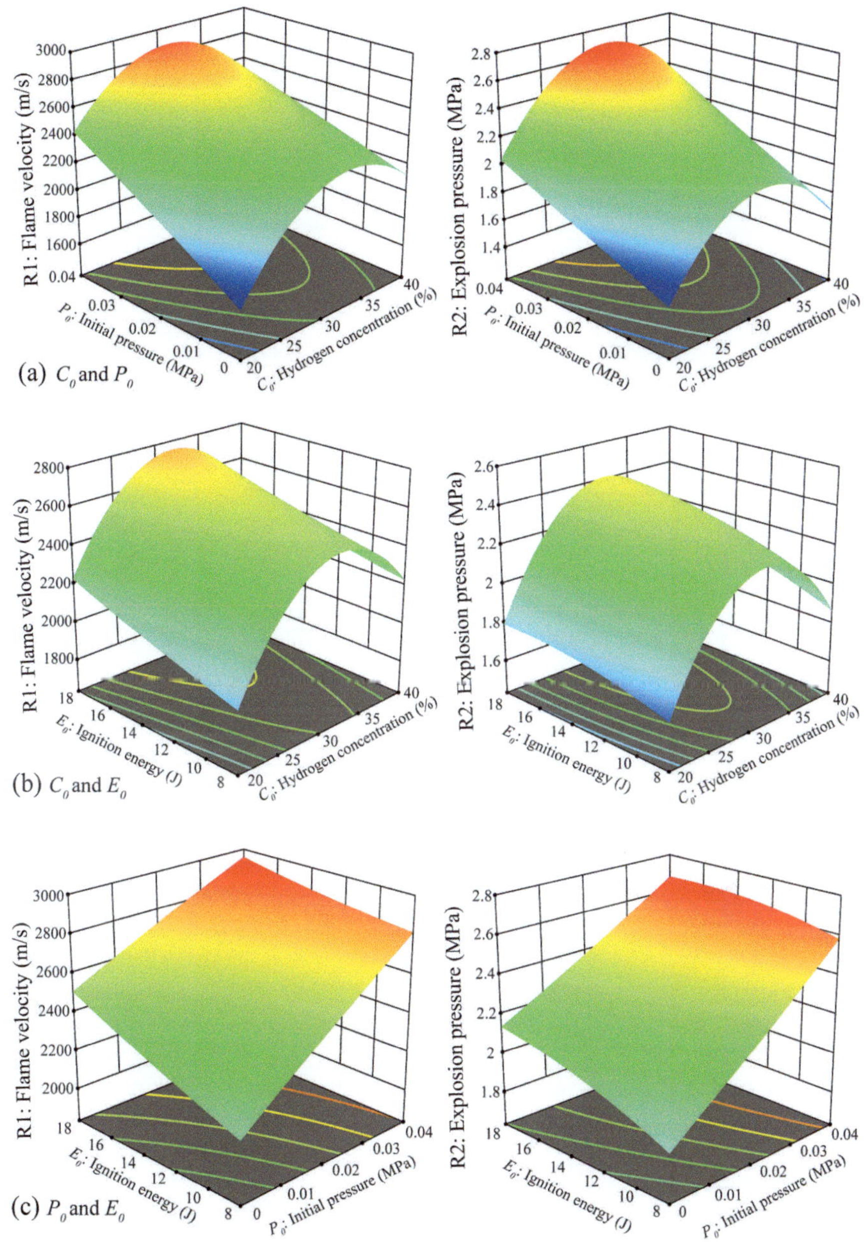

Fig. 12.7 The influence diagrams of the two-factor interaction (**a** C_0 and P_0; **b** C_0 and E_0; **c** P_0 and E_0)

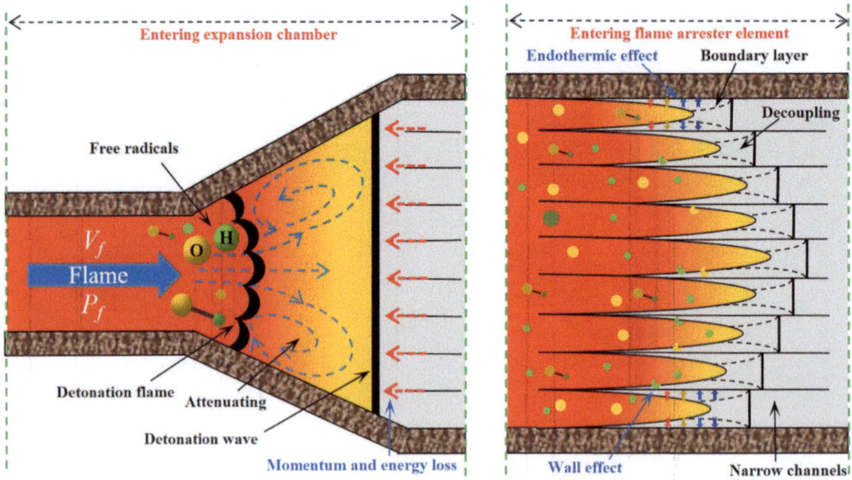

Fig. 12.8 Sketch illustrating detonation quenched in the EC and FA-E

flame temperature reduces the quenching critical value. Then, changes in the structure parameters of FA-E can also influence the attenuation effects such as the endothermic and the wall effects inside the FA, thus influencing the quenching performance of the FA. The collision effect between the detonation and the front end face of the FA-E is enhanced as the porosity decreases, leading to the greater increases in the velocity deficit. Besides, the flame will also be divided into tinier flames, and the contact area between the flame and the surface of channel wall is increased. And the increases of the wall endothermic, the boundary layer and the free radical destruction effects cause the decoupling and quenching of the detonation. The action time between flame and the surface of channel wall is extended due to the rise in element thickness, which enhances the attenuation effect of narrow channel on detonation and leads to the reduction in the flame velocity. Finally, the detonation is successfully decoupled and quenched.

In addition, initial conditions indirectly impact the quenching performance of FA by influencing the V_f and P_f accessing the FA [13]. The propagation time of flame inside EC and narrow channel is greatly reduced due to the higher velocity and pressure of the detonation [8]. This will reduce the endothermic and wall effects of FA on detonation, leading to the reduction in velocity deficit and the flame being in a high-velocity propagation state inside FA. Hence, the flame temperature and the free radicals number do not reach the critical value for flame quenching, and the flame continues to spread. In summary, the quenching effect of the FA is affected by the initial conditions and the structure parameters of the FA and EC. These factors directly affect the V_f and P_f accessing the FA, and the process of the endothermic and the wall effects, thus indirectly affecting the quenching performance of FA.

12.3 Conclusions

The experimental study on the quenching performance and influence law of crimped-ribbon flame arrester (CFA) on hydrogen detonation was carried out. The influence of initial conditions, structure parameters of the flame arrester (FA) and expansion chamber (EC) on the explosion parameters, the attenuation extent and quenching results of the FA were studied. Meanwhile, the multi-factor prediction models for V_f and P_f under the action of initial conditions were established separately by RSM. The conclusions were as follows:

(1) The initial conditions indirectly influenced the quenching performance by affecting the V_f and P_f. Both V_f and P_f attained the maximum values at the hydrogen concentration of 30%, and the attenuation effect of the FA on the detonation velocity was minimized. The V_f and P_f were significantly diminished as the hydrogen concentration was far away from 30%, and the attenuation extent of FA was enhanced. Meanwhile, both V_f and P_f values were raised by increasing initial pressure and ignition energy, and the attenuation extent was gradually weakened. Furthermore, the diminution of porosity and the rise of element thickness would significantly improve the attenuation effect of FA on detonation, and the porosity had a better influence on the quenching performance.

(2) The detonation intensity acting on the flame arrester element (FA-E) was reduced due to the attenuation effect of the EC, and further affecting the quenching results of FA. The attenuation effect of the ECs on flame velocity was minimized at the hydrogen concentration of 30%, and the FA with conventional and baffle type EC failed. Meanwhile, the baffle and extended type EC were more effective than the conventional type EC in reducing the detonation velocity entering the FA-E, and the quenching performance of extended type EC was better. Furthermore, the multi-factor prediction models of V_f and P_f under the action of initial conditions were built by RSM. The influence order of single factor on V_f and P_f was: $P_0 > C_0 > E_0$; and the influence order of the two-factors interaction was: (C_0 and P_0) > (P_0 and E_0) > (C_0 and E_0).

(3) The quenching of detonation by CFA was largely ascribed to the attenuation effect of the EC, the endothermic and the wall effects of the FA-E. Meanwhile, the attenuation effects of the FA would be reduced as the detonation had higher velocity and pressure, which decreased the velocity deficit and caused the flame to be difficult to quench. The structure change of the EC could enhance the momentum and energy loss, leading to the detonation being more easily quenched by narrow channel. In addition, the increase of the element thickness and the diminution of the porosity of the FA-E would enhance the wall endothermic and the free radical destruction effect, which increased the flame velocity deficit and caused the detonation to be successfully quenched.

For the industrial practice, three guidance suggestions are proposed to improve the performance of CFA on hydrogen detonation and realize its optimal design: (1)

To reduce the detonation intensity and avoid its occurrence, the transmission pressure in the pipeline should be decreased as much as possible under the condition of satisfying the gas transmission. (2) The structure should be further optimized based on the extended EC to weaken the detonation intensity acting on the FA-E. (3) To improve the performance of FA for quenching highly-active hydrogen detonation, the influence of porosity should be prioritized. The study on the quenching of hydrogen detonation by crimped-ribbon flame arrester obtains the influence laws and prediction models of different factors. This is of great significance for the optimization design of industrial explosion inhibition equipment and the improvement of explosion prevention theory, and provides safety guarantees for the transportation of hydrogen pipelines and industrial production.

References

1. Lin CD, Cao XY, Wang ZR, Weo JS, Xu JJ, Sun SC. Multi-factor influencing on detonation quenching performance of hydrogen. J Loss Prev Process Ind. 2024;87:105236.
2. Yan W, Wang Z, Ouyang D, et al. Analysis and prediction of thermal runaway propagation interval in confined space based on response surface methodology and artificial neural network. J Energy Storage. 2022;55:105822.
3. Rodriguez V, Jourdain C, Vidal P, et al. An experimental evidence of steadily-rotating overdriven detonation. Combust Flame. 2019;202:132–42.
4. Pan Z, Qi J, Pan J, et al. Fabrication of a helical detonation channel: effect of initial pressure on the detonation propagation modes of ethylene/oxygen mixtures. Combust Flame. 2018;192:1–9.
5. Bao L, Wang P, Dang W, et al. Experimental study on detonation flame penetrating through flame arrester. J Loss Prev Process Ind. 2021;72:104529.
6. Chu J, Liu Y, Lv X, et al. Experimental investigation on blockage predictions in gas pipelines using the pressure pulse wave method. Energy. 2021;230:120897.
7. Sun SC, Shu Y, Feng Y, et al. Numerical simulation of detonation wave propagation and quenching process in in-line crimped-ribbon flame arrester. Cogent Eng. 2018;5:1469377.
8. Chao J, Ng HD, Lee JHS. Detonability limits in thin annular channels. Proc Combust Inst. 2009;32:2349–54.
9. Kundu S, Zanganeh J, Moghtaderi B. A review on understanding explosions from methane-air mixture. J Loss Prevent Proc. 2016;40:507–23.
10. Chen X, Zhao N, Zheng H, et al. Effect of ignition parameters on detonation initiation using toroidal shock wave focusing. Aero Sci Technol. 2021;109:106421.
11. Zhang Y, Li C, Chu D, et al. Process optimization for the preparation of thiamethoxam microspheres by response surface methodology. React Funct Polym. 2020;147:104460.
12. Zhou KY, Li ZF. The attenuation of gaseous detonation strength as a result of the bumper-separator in expanding cavity of the detonation arrester. Explos Shock Waves. 2001;21:179–83.
13. Zhang B, Liu H, Yan B. Investigation on the detonation propagation limit criterion for methane-oxygen mixtures in tubes with different scales. Fuel. 2019;239:617–22.

MIX
Papier aus verantwortungsvollen Quellen
Paper from responsible sources
FSC® C105338

If you have any concerns about our products,
you can contact us on
ProductSafety@springernature.com

In case Publisher is established outside the EU,
the EU authorized representative is:
Springer Nature Customer Service Center GmbH
Europaplatz 3, 69115 Heidelberg, Germany

Printed by Libri Plureos GmbH
in Hamburg, Germany